T0351017

Sustainable Automated Production Systems

Due to disruptions occurring within production processes as a result of different circumstances, it is necessary to provide a new and safe production system that can be resilient to cope with any crisis, such as a pandemic. New sustainable automated production systems need to be proposed and adopted by industrial managers and this book fulfills that need.

Sustainable Automated Production Systems: Industry 4.0 Models and Techniques includes the modern technologies of Industry 4.0 and discusses automated systems, robotic production, sustainable automation, production process automation, and the use of IoT in production. It goes over the circular economy and automated production processes and presents theories and practices associated with mathematical models of process automation in sustainable production. The book also investigates how sustainability in automated production systems using the IoT can present unique circumstances when pandemics or other disruptions are involved.

The book includes chapters presenting sustainable production models and automated production using Industry 4.0 technologies and is targeted to help both practitioners and academics.

Sustainable Automated Production Systems

Industry 4.0 Models and Techniques

Hamed Fazlollahtabar

CRC Press
Taylor & Francis Group
Boca Raton London New York

CRC Press is an imprint of the
Taylor & Francis Group, an **informa** business

Designed cover image: Shutterstock–VectorMine

MATLAB® and Simulink® are trademarks of The MathWorks, Inc. and are used with permission. The MathWorks does not warrant the accuracy of the text or exercises in this book. This book's use or discussion of MATLAB® or Simulink® software or related products does not constitute endorsement or sponsorship by The MathWorks of a particular pedagogical approach or particular use of the MATLAB® and Simulink® software.

First edition published [2024]
by CRC Press
2385 NW Executive Center Drive, Suite 320, Boca Raton FL 33431

and by CRC Press
4 Park Square, Milton Park, Abingdon, Oxon, OX14 4RN

CRC Press is an imprint of Taylor & Francis Group, LLC

© 2024 Hamed Fazlollahtabar

Reasonable efforts have been made to publish reliable data and information, but the author and publisher cannot assume responsibility for the validity of all materials or the consequences of their use. The authors and publishers have attempted to trace the copyright holders of all material reproduced in this publication and apologize to copyright holders if permission to publish in this form has not been obtained. If any copyright material has not been acknowledged please write and let us know so we may rectify this in any future reprint.

Except as permitted under U.S. Copyright Law, no part of this book may be reprinted, reproduced, transmitted, or utilized in any form by any electronic, mechanical, or other means, now known or hereafter invented, including photocopying, microfilming, and recording, or in any information storage or retrieval system, without written permission from the publishers.

For permission to photocopy or use material electronically from this work, access www.copyright.com or contact the Copyright Clearance Center, Inc. (CCC), 222 Rosewood Drive, Danvers, MA 01923, 978-750-8400. For works that are not available on CCC please contact mpkbookspermissions@tandf.co.uk

Trademark notice: Product or corporate names may be trademarks or registered trademarks and are used only for identification and explanation without intent to infringe.

ISBN: 978-1-032-50576-3 (hbk)
ISBN: 978-1-032-50988-4 (pbk)
ISBN: 978-1-003-40058-5 (ebk)

DOI: 10.1201/9781003400585

Typeset in Times
by Newgen Publishing UK

Contents

Preface

Sustainability is a term that now permeates, with all its nuances, our workplaces, homes, schools and many other organized groups, as a true pop star, bringing a sense of responsibility and a Utopian, or not, search for a global society, conscious of its actions and interdependence. The speed of adoption of sustainable solutions is well below the desired, as it requires policy makers to have the will to force changes; business will to create significant improvements in their means of production in addition to the determined pressure by society as a whole. Society, similarly, must adopt the awareness of sustainability and avoid uncontrolled consumerism.

Production automation often referred to as factory automation or industrial automation uses a combination of software, computers, and robotics (or control systems) in place of human labor. Economic and global shifts affect production trends and the changing needs and evolution of Industry 4.0 are matched by changes in production automation. As the world becomes more acutely aware of the effects of climate change, demands for more sustainable production practices are on the rise. Production automation companies have helped the industry at large keep pace with this demand. Automated production is inherently sustainable. Its existence allows companies to create and use processes that minimize negative environmental impact. This helps companies to achieve sustainable production goals including lowered energy, water, and land usage.

AGVs have been in use for over 50 years. The first AGV was built and introduced in 1953 at a grocery warehouse. It was a driverless towing truck, which followed an overhead wire. After a few years, many types of warehouses and factories had started to use towing AGVs and still do. In 1973, Volvo in Sweden set out to develop an alternative to the conventional assembly line, and they achieved this by introducing 280 AGVs. The next step in the AGV evolution was the unit load AGV and when it was introduces in the mid 70's it gained widespread acceptance because of its ability to serve multiple functions, such as transportation device and assembly platforms. After the 70's different guiding systems were introduced which increased the flexibility of the AGVs, for example, laser guidance. The AGV system continued to evolve at the same pace as the advances within electronics and computers, and today there are over 15 different types of AGVs with around 40 manufactures worldwide.

Effective design of automatic material handling devices is one of the most important decisions in flexible cellular manufacturing systems. Minimization of material handling operations could lead to optimization of overall operational costs. An AGV is a driverless vehicle used for the transportation of materials within a production plant partitioned into cells. The tandem layout is according to dividing workstations into non-overlapping closed zones to which a Tandem Automated Guided Vehicle (TAGV) is allocated for internal transfers. Shirazi et al. (2010) illustrated a non-linear multi-objective problem for minimizing the material flow intra and inter-loops and minimization of the maximum amount of inter cell flow, considering the limitation of TAGV work-loading. For reducing variability of material flow and establishing balanced zone layout, some new constraints have been added to the problem based on the Six-Sigma approach. Due to the complexity of the machine grouping control problem, a modified ant colony optimization algorithm was used for solving this model. Fazlollahtabar et al. (2015) were concerned with both time and cost optimization in an automated manufacturing system. In their proposed system, AGVs and robots for the manufacturing functions carry out material handling. In this way, defect rates, breakdown times, waiting times, processing times, and certain other parameters are not expected to be deterministic. Stochastic programming was applied to optimize the production time and material handling cost. The proposed stochastic program was a nonlinear model so for this reason a Successive Linear Programming (SLP) technique was employed for its optimization. Numerical test results pointed to the inefficacy of the proposed optimization method for large sized problems. Hence, a Genetic Algorithm (GA) was presented to optimize large sized problems.

The optimization of Material Handling Systems (MHSs) can lead to substantial cost reductions in manufacturing systems. Choosing adequate and relevant performance measures is critical in accurately evaluating MHSs. The majority of performance measures used in MHSs are time-based. However, moving materials within a manufacturing system utilizes time and cost. Tavana et al. (2014) considered both time and cost measures in an optimization model used to evaluate an MHS with AGVs. They considered the reliability of the MHSs because of the need for steadiness and stability in the automated manufacturing systems. Reliability was included in the model as a cost function. Fazlollahtabar and Shafieian (2014) were concerned with the design of a computer integrated manufacturing system to identify an optimal path in a Vehicle Routing Problem (VRP) network with respect to triple criteria. In most VRPs just one criterion, either time or cost was considered in decision-making. They considered all time, cost, and AGV capability in decision making, simultaneously. To satisfy a Material Requirement Planning (MRP) by providing the Bill Of Material (BOM), AGVs are suitable devices. Fazlollahtabar et al. (2012) were concerned with applying Tandem Automated Guided Vehicle (TAGV) configurations as material handling devices and optimizing the production time considering the effective time parameters in a Flexible Automated Manufacturing System (FAMS) using the Monte Carlo simulation. Due to different configurations of TAGVs in an FAMS, the material handling activities are performed. With respect to various stochastic time parameters and the TAGV defects during material handling processes, sample data was collected and their corresponding probability distributions were fitted. Using the probability distributions, they modelled the TAGV material handling problem via the Monte Carlo simulation. The effectiveness of the proposed model was illustrated in a case study.

An Automated Manufacturing System (AMS) is a complex network of processing, inspecting, and buffering nodes connected by system of transportation mechanisms. For an AMS, it is desirable to be capable of increasing or decreasing the output with the rise and fall of demand. Such specifications show the complexity of decision making in the field of AMSs and the need for concise and accurate modeling methods. Therefore, Fazlollahtabar et al. (2010) proposed a flexible job shop automated manufacturing system to optimize material flow. The flexibility was on the multi-shops of the same type and multiple products that can be produced. An automated guided vehicle was applied for material handling. The objective was to optimize the material flow regarding the demand fluctuations and machine specifications. Fazlollahtabar and Mahdavi-Amiri (2013) proposed an approach for finding an optimal path in a flexible job shop manufacturing system considering two criteria of time and cost. A network was configured in which the nodes are considered the shops with arcs representing the paths among the shops. An automated guided vehicle functioned as a material handling device through the manufacturing network. The expert system for cost estimation was based on fuzzy rule backpropagation network to configure the rules for estimating the cost under uncertainty. A multiple linear regression model was applied to analyze the rules and find the effective rules for cost estimation. The objective was to find a path minimizing an aggregate weighted unscaled time and cost criteria. A fuzzy dynamic programming approach was presented for computing the shortest path in the network. Then, a comprehensive economic and reliability analysis was worked out on the obtained paths to find the optimal producer's behavior. Fazlollahtabar and Mahdavi-Amiri (2013) proposed an approach for finding an optimal path in a flexible job shop manufacturing system considering two criteria of time and cost. With rise in demands, advancement in technology and increase in production capacity, the need for more shops persists. Therefore, a flexible job shop system has more than one shop with the same duty. The difference among shops with the same duty is in their machines with various specifications. A network was configured in which the nodes were considered the shops with arcs representing the paths among the shops. An Automated Guided Vehicle (AGV) functioned as a material handling device through the manufacturing network. To account for uncertainty, the authors considered time to be a triangular fuzzy number and applied an expert system to infer cost. The objective was to find a path minimizing both the time and cost criteria, aggregately. Since time and cost have different scales, a normalization procedure was proposed to remove the scales. The model being biobjective, the analytical hierarchy process weighing method was applied to construct a single objective. Finally, a dynamic programming approach was presented for computing a shortest path in the network. The efficiency of the proposed approach was illustrated by a numerical example. Fazlollahtabar and Mahdavi-Amiri (2013) proposed a cost estimation model based on a fuzzy rule backpropagation network, configuring the rules to estimate the cost under uncertainty. A multiple linear regression analysis was applied to analyze the rules and identify the effective rules for cost estimation. Then, using a dynamic programming approach, they determined the optimal path for the manufacturing network. Finally, an application of the model was illustrated through a numerical example showing the effectiveness of the proposed model for solving the cost estimation problem under uncertainty. Fazlollahtabar and Olya (2013) concerned with proposing a heuristic statistical technique to compute total stochastic material handling time in an Automated Guided Vehicle (AGV) equipped job shop manufacturing system. With respect to stochastic times of AGVs material handling process, the material handling activities probability distributions were considered. Using the probability distributions, they modeled the AGV material handling problem using a heuristic statistical method when the activities' probability distribution functions were the same. Also, in the case that the activities' probability distribution functions were different, a cross-entropy approach was proposed and developed to model the problem. The effectiveness of the proposed model was illustrated in a numerical example and verified by a simulation study.

Some of the ways that production automation helps the environment are that data collected is used by manufacturers to ensure their machines are efficient; more efficient machines have lower heating requirements; robots help to streamline processes, meaning less material waste; automation allows for reduced cycle times, resulting in reduced energy output; robots can be compact and can streamline processes thus requiring less factory floor space. From a business perspective, production automation can help companies reach Key Performance Indicators (KPIs) related to corporate sustainability and environmental regulatory compliance. A combination of regulations and internal guidelines has resulted in industrial companies seeing cost savings and increased efficiency. Currently, facing pandemics in productions systems is a rising challenge to keep sustainability and the demand side management, simultaneously. Industry 4.0 tools could help to handle production management level decisions using the Internet of Things (IoT). In this book, sustainability in automated production systems using the IoT is investigated with a focus on pandemic circumstances.

REFERENCES

Fazlollahtabar, H. and Mahdavi-Amiri, N. (2013). An optimal path in a bi-criteria AGV-based flexible job shop manufacturing system having uncertain parameters. *International Journal of Industrial and Systems Engineering*, 13(1), 27–55.

Fazlollahtabar, H. and Olya, M.H. (2013). A cross-entropy heuristic statistical modeling for determining total stochastic material handling time. *The International Journal of Advanced Manufacturing Technology* 67, 1631–1641.

Fazlollahtabar, H. and Shafieian, S.H. (2014). An optimal path in an AGV-based manufacturing system with intelligent agents. *Journal for Manufacturing Science and Production* 14(2), 87–102.

Fazlollahtabar, H., Mahdavi-Amiri, N. and Muhammadzadeh, A. (2015). A genetic optimization algorithm for nonlinear stochastic programs in an automated manufacturing system. *Journal of Intelligent & Fuzzy Systems* 28(3), 1461–1475.

Fazlollahtabar, H., Es'haghzadeh, A., Hajmohammadi, H. et al. (2012). A Monte Carlo simulation to estimate TAGV production time in a stochastic flexible automated manufacturing system: a case study. *International Journal of Industrial and Systems Engineering* 12(3), 243–258.

Fazlollahtabar, H., Rezaie, B. and Kalantari, H. (2010). Mathematical programming approach to optimize material flow in an AGV-based flexible jobshop manufacturing system with performance analysis. *The International Journal of Advanced Manufacturing Technology* 51(9–12), 1149–1158.

About the Author

Hamed Fazlollahtabar earned a BSc and an MSc in Industrial Engineering from Mazandaran University of Science and Technology, Iran, in 2008 and 2010, respectively. He received his PhD in Industrial and Systems Engineering from Iran University of Science. He has completed a postdoctoral research fellowship at Sharif University of Technology, Tehran, Iran, in the area of reliability engineering for complex systems from October 2016 to March 2017. He joined the Department of Industrial Engineering at Damghan University, Damghan, Iran, in June 2017 and currently is an associate professor of Industrial Engineering. He has been listed in the top 2% of scientists in the world in 2020, 2021 and 2022. He has been selected as the best researcher of all engineering disciplines in Iran 2022. He has been selected as the distinguished young researcher of Industrial Engineering in Iran 2023. He is on the editorial boards of journals and technical committees of conferences. His research interests are in robotic production systems, reliability engineering, sustainable supply chain planning, and business intelligence and analytics. He has published more than 300 research papers in international books, journals, and conferences. He has also published ten books out of which seven are internationally distributed to academics.

1 Sustainability in Production

1.1 INTRODUCTION

Based primarily on offshore and centralized facilities with large scale assembly lines to supply a mass market, the current manufacturing model is driven to change, by new technologies promoted in visions such as Industry 4.0, into a decentralized, on demand, localized, and customizable manufacturing model known as Re-distributed Manufacturing (RdM). The UK Engineering and Physical Sciences Research Council (EPSRC) have a working definition of Re-Distributed Manufacturing (RdM) as "Technology, systems, and strategies that change the economics and organization of manufacturing, particularly with regard to location and scale". Moreno and Charnley (2016) corroborate this definition and state that RdM "enables a connected, localized, and inclusive model of consumer goods production and consumption that is driven by the exponential growth and embedded value of big data". This new model aims to apply Industry 4.0 technologies to help change the organization of manufacturing in terms of location and scale in order to reduce supply chain costs, improve sustainability, and provide customizable products more akin to an individual customer's needs. Industry 4.0 is an initiative between the German government[1] and national industries to envisage and promote the use of new technologies and organizational methods for manufacturing.

The advent of this new manufacturing paradigm has brought on the need for models and methods that manufacturers can rely on as guides for the implementation of RdM processes into their operations. The use of emerging technologies in the design and manufacture of consumer goods (Manyika et al., 2015) such as information communication technologies, automation and robotics, big data analytics, additive manufacturing, cloud computing, and mobile technologies could enable intelligent and digitally networked manufacturing systems, resulting in the redistribution of manufacturing towards smaller scale manufacturing processes (Freeman et al., 2017).

A case study is used to develop an initial distributed and circular business model. The Circular Economy concept aims to promote a move from the current take-make-dispose model of production to one that minimizes waste through improved product design and encourages reuse and recycling of materials (MacArthur, 2013; Stahel, 2016; Geissdoerfer et al., 2017). The selected case study was drawn from the Shoe4.0 project, a collaboration which aims to develop a proof of concept for a smart and

DOI: 10.1201/9781003400585-1

1

sustainable shoe. This case study by Yin (2013) was perceived as the most suitable for answering the research question of: How could we develop a re-distributed and circular business model? The developed business model was used to investigate how data captured from, and communicated between, supply, production, distribution, and use can be used to design improved processes.

As part of 'the Shoe4.0' model, a pair of 3D-printed trainers were designed and prototyped by considering circular economy aspects such as the elimination of waste through the efficient use of materials and production technologies, the removal of toxic chemicals that impair reuse, a reduction in manufacturing processes and parts use, and design for disassembly (to enable maintenance, reuse, and refurbishment of produced goods). A Selective Laser Sintering (SLS) technology was selected with a Thermoplastic Polyurethane Elastomer (e.g., Duraform Flex) as the base material to produce the trainers. This material is fully recyclable and can be used again in a SLS printer. SLS is a form of Additive Manufacturing (also known as 3D printing), a production technique that utilises digital designs to be created in 3D physical form through layer-by-layer deposition of material (Ford and Despeisse, 2017). Its properties are ideal for the footwear industry as it is flexible, durable, tear-resistant, soft-touch, and washable. The design of this pair of trainers allows new disruptive business models, such as offering trainers as a service through a subscription model. This model provides a personalized service if the trainers need to be repaired, maintained, or parts need to be replaced, as the main body detaches from the sole with a mechanical joint. In addition, trainers will be produced in local stores. The model also includes the use of other technologies such as the ability to scan your foot to produce every trainer to measure and an augmented reality application to virtually try the trainers on. These technologies will allow the custom production of trainers avoiding a surplus of unsold products and utilizing the minimal amount of material.

This study aims to explore through IDEF0 (Icam DEFinition for Function Modelling) the viability of a new business model for manufacturers employing additive manufacturing processes, such as 3D printing, as part of a circular and re-distributed production and consumption model for the footwear industry. A review of existing literature will reveal whether there are business models designed using IDEF0 that can directly support the implementation of a re-distributed way of manufacturing. In particular, models containing parameters concerned with transportation, customer involvement, servitisation, and circularity will be sought. In seeking to establish if such models for circular and redistributed production exist, this chapter will scope an agenda by putting forward new models derived from the Shoe4.0 case study.

To achieve this aim, a five-step method of the IDEF0 Model, presented in Section 4, was used to model an As-Is value chain as well as a new business model to make a comparison of traditional and re-distributed manufacture. To develop the IDEF0 model, a literature review was conducted to establish the state-of-the-art in re-distributed manufacturing and its constituent technologies and existing business models (Section 2). Both Shoe4.0 models and their validation are further explained in Section 4. Section 5 discusses the reasoning behind the selection of IDEF0 as a tool and its application in modelling a consumer goods manufacturing As-Is value chain, how criteria were developed in order to determine the inputs, outputs, resources, and controls terminology to be used by each model, and a comparison of both models revealing some important insights about the challenges of implementing re-distributed models of production and consumption in the consumer goods sector. Finally, the chapter concludes in Section 6 by giving some recommendations on how a RdM model could thrive within this industry.

1.2 RELATED WORKS

Underpinning the RdM movement are a number of enabling concepts such as servitisation, customization, and localization (Moreno et al., 2019). This literature review research was conducted on these three main subjects for RdM and further expanded upon by exploring concepts such as Circular Economy, and manufacturing processes in more depth, alongside the related technology espoused in Industry 4.0 initiatives.

In terms of servitisation, this concept is exemplified by Baines et al. (2007) as the "... customer pays for using an asset, rather than its purchase, and so benefits from a restructuring of the risks, responsibilities, and costs traditionally associated with ownership." This means that the line between what's considered a product and a service could be blurred into a product-service system which would help manufacturers "... sustain competitiveness, ... 'move up the value chain' and (deliver) knowledge intensive products and services."

While helping to develop and maintain customer loyalty by providing more complete offerings through a product-service system, Kastalli and Van Looy (2013) mention that the choice of business model for a manufacturer turned service provider, and their managerial practices, are crucial to successfully create products that can be complemented by services and vice versa. Providing solutions through a combination of products and services implies more participation between the manufacturer and customer during the product's life. The focus could then be placed on providing a purposeful service model that would mostly make "manufacturers or retailers retain ownership of their products (or have an effective take-back arrangement) and, where possible, act as service providers, selling the use or performance of products, not their consumption." (Moreno et al., 2017). According to Moreno and Charnley (2016), benefits from the implementation of RdM in the consumer goods sector could bring ways to "... effectively manage resources within markets, ensure waste is eliminated and monetized (Lacy and Rutqvist, 2015), and support selling products as services, which will enable keeping products in longer use to minimize waste and resources".

This new paradigm of manufacturer-customer relationship and product/materials use and re-use is known as a Circular Economy (CE). A CE draws attention to the entire life cycle of the materials from the moment they are sourced raw from suppliers to when they are transformed by manufacturers, dispensed by distributors, and sold by retailers, to the end consumer for use; even further into considering the reverse supply chain, in which expended products are used as new input to initiate another cycle in what is understood as a closed-loop supply chain system (Lacy and Rutqvist, 2015. In this way, CE aims to increase the efficiency of resource use, with a special focus on urban and industrial waste, to achieve a better balance and harmony between economy, environment, and society. This considers the technological, societal, and environmental aspects in terms of individual industrial processes (Ghisellini et al., 2016).

Being able to take products that would be considered at the end of their useful life and give them a new purpose (reuse) can be very beneficial in many ways. There are obvious environmental and economic benefits from requiring fewer resources and therefore, less energy and less labor compared to when a product is being created from raw materials, recycled, or disposed. Ghisellini et al. (2016) continue to explain that if the reuse of products is to be propagated then products themselves must be designed to be durable and prepared for multiple use cycles, coupled with incentives from companies to encourage take-back of products.

The customer is constantly cultivating and refining their tastes and requirements for novel products. This need from the customer can be satisfied by manufacturers by providing a service of mass customization (Hasan et al., 2013). Mass Customization (MC) can be defined as the ability to manufacture tailor-made personalized products and services utilizing technologies and systems with near mass production costs and efficiency (Boeer et al., 2013).

There are some substantial hurdles to overcome in the application of mass customization models. Currently, the extent to which customization can be completed is limited by the manufacturers' particular production capabilities, competitive positions, and by the technology and methods available to them. Technology can currently only manage to personalize a limited amount of attributes and products (Mai et al., 2016). Therefore, any manufacturer wishing to implement a mass customization model should consider their "ability to deliver on and integrate the three strategic elements: elicitation, process flexibility, and logistics. Thus, highly flexible production technologies such as 3D printing can be used to respond quickly to customers' wants and needs".

3D printing is a type of additive manufacturing in which an item is built by adding layer after layer of its component material in a successive process until a three dimensional object is created; "which differs from the more usual "subtractive" (when an object is carved out of a block of raw material) or moulding/die-casting (when a molten material is injected into a solid mould) forms of manufacturing" (Rayna and Striukova, 2016). In addition to enabling the production of tailor-made personalized products, 3D printing allows for design and manufacturing ideas to be tried and tested at significantly greater speeds, consequently improving the rate at which product innovation occurs.

In addition, concepts such as Industry 4.0, a term that comprises the horizontal integration of the value creation network created by emerging technologies and applications (like additive and cloud manufacturing), can help to create an end-to-end commerce system that covers the entire product life cycle with a network of distributed manufacturing arrangements. This type of facility will be "smart," in the sense of being composed of networks of intelligent processes that allow the product to dictate when it will be produced and also what machine or group of machines will carry out the process and what would be their efficient utilization in meeting the requested quantities (Matt et al., 2015). Therefore, concepts such as servitisation, customization, and localization, alongside the introduction of new technologies, give rise to what is now known as RdM manufacturing. RdM can be defined as "… an emerging concept which captures the anticipated reshoring and localization of production from large-scale mass manufacturing plants to smaller-scale localized, customizable production units, largely driven by new digital production technologies" (Pendeville et al., 2016).

A successful implementation of this new manufacturing paradigm will combine all the new technologies and methods into a coherent system. The best way of achieving this is by first creating suitable business models to follow. This new "… business model represents a new subject of innovation, which complements the traditional subjects of process, product, and organizational innovation and involves new forms of cooperation and collaboration" (Zott et al., 2011). Since new technologies like additive manufacturing must be integrated into the current processes this "can open up new subspaces in the existing technical performance and functionality space, which in turn requires a new business model if the economic value potential of the new technology is to be captured." New business models will realize the benefits from all novel commercial offerings that the implementation of a new manufacturing paradigm will create.

Matt et al. (2015) provide a classification of eight different models for RdM, which provides a complete spectrum from the most basic to the most visionary. Manufacturing facilities of Type One will apply the most elementary concepts of RdM by replicating the same standardized procedures in geographically dispersed facilities. Manufacturing facilities of Type Four would incorporate a degree of flexibility to adapt to their surroundings. This would come in the form of changes in customer trends and environmental changes. This would be achieved by implementing digital network technology that would enable the product itself to dictate how many units, with what specifications, and the configuration of machines required to produce it. Manufacturing facilities of Type Eight would represent the most innovative model using cloud manufacturing and additive manufacturing, selling product data and bringing the physical production to within close proximity of the customer.

Based on the review of existing literature, it is evident that there are no business models designed using IDEF0 that can directly support the implementation of a re-distributed way of manufacturing. In particular, parameters concerned with transportation, customer involvement, servitisation, and circularity are lacking in published models. If a business in the manufacturing industry is interested in converting to a distributed and sustainable way of production, then the lack of proper business models to implement can be a hindrance to that process. Coupled with a general lack of guidance in how to approach the implementation of RdM within an organisation, these elements are the focus of the research presented in this chapter.

1.3 METHODOLOGY

The IDEF0 (Icam DEFinition for Function Modelling) modelling standard is one of the most popular graphical notations for business systems and process planning (Sugiyama and Hirao, 2014; Mu et al., 2015). IDEF0 could provide a clear picture of

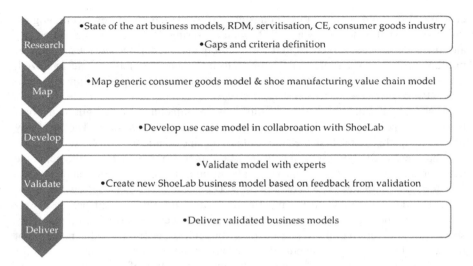

FIGURE 1.1   Proposed Methodology.

how value can be created and then transferred through the help of different functions and resources during the implementation of a distributed and circular business model. Such a model requires a systemic perspective in order to connect and deliver value to the region where it should integrate in order to be meaningful and provide value as expected. This study has followed a linear development path involving the establishment of the state-of-the-art through value chain mapping in the consumer goods industry to the development of a business model for sustainable RdM based on a case study (shown in Figure 1.1). A generalized manufacturing value chain serves as a foundation for all of the models as they follow a procure, produce, retail, consume, and dispose flow of functions as a structure/initial blueprint. The case study as a methodological tool for intensive investigation into one research case instance is now well established (Gerring, 2016). The case study development process, shown in Figure 1.2, involved three broad stages: brainstorming within the team for development of initial scenario areas, literature review and research gap identification; development of As-Is and To-Be scenarios, feedback on initial scenarios from Shoe4.0 internal and steering groups; validation of completed Integration Definition for Function Modelling (IDEF) model scenarios by experts, and re-evaluation of case study models. The main procedures for the development of the models were followed according to Integration Definition for Function Modelling (IDEF)[2] model creation methodology.

Prior to the development of the models, a value chain approach helped determine the main functions that each of the models would roughly contain, providing a general flow to define inputs and outputs from one function to the next. Furthermore, criteria were developed based on the understanding of the subjects in the literature review in order to provide guidance for the definition of controls and resources

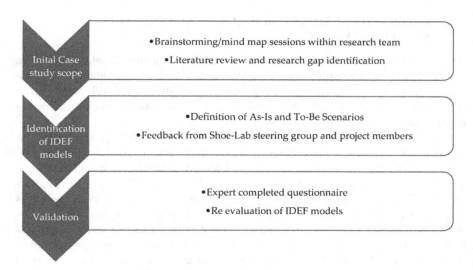

FIGURE 1.2   Case Study Development Process.

that each function would contain. These criteria consisted of parameters identified while developing the models around four major concepts dealing with transportation, customer involvement, servitisation, and circularity. On development of the Shoe4.0 case study, an As-Is was created specific to the shoe manufacturing process in order to have a more accurate reference model to compare against the completed case study business model. During the case study development stage, information was gathered about the intelligent shoe concept through a variety of meetings with experts developing the Shoe4.0 project. These meetings consisted of presentations made to Shoe4.0 project and steering group members as well as two different focus groups, during which information regarding the concept was further explored with participants who were not directly involved in the project. The focus groups took the form of professionally facilitated and led discussions utilizing brainstorming, idea wall/post it notes, and mind mapping tools Once sufficient knowledge on the subject was acquired, a draft model of the Shoe4.0 value chain was created.

The Shoe4.0 model was then validated through the creation of a closed question survey with open comment sections. The survey was completed by ten experts drawn from both industry and academia. The feedback recovered from the survey answers was then utilised to further improve the Shoe4.0 case study model. A new Shoe4.0 concept model based on the knowledge gathered throughout the project was then created in order to explore other possible variations on the RdM models already created. This would allow further contrast between the As-Is shoe manufacturing process model, the Shoe4.0 case study, and concept models. All models produced at this stage were compared in order to gather insights into the possible contributions that the models could provide to the future development of RdM models and their implementation.

1.4 PROPOSED BUSINESS MODEL FOR SUSTAINABLE MANUFACTURING

As previously mentioned, the case study is drawn from the Shoe4.0 case study. Shoe4.0 was a collaboration to improve the current model for designing, manufacturing, distributing, selling, using, and disposing of shoes by developing a novel business model based on RdM and circular economy concepts to provide an enhanced product-service offering to the customer.

The concept is based on a modular shoe design which is made up of few components, allowing a more efficient use of materials, reparability, and upgradability. These components can be separated at end-of-life for re-use or recycling. This is enabled by the use of 3D scanning and printing technology to manufacture the different components of the shoe according to the specific customer's foot dimensions, intended product use (serious runner, casual runner, fashion) and aesthetic preferences. The shoes will be embedded with sensors in order to provide benefits to the customer such as performance monitoring, location tracking, and shoe condition depending on the customer's intended use and preferences. Depending on the type of sensor used, the technology could provide information on how the wearer moves and when the shoe needs to be repaired or replaced and additionally, inform the manufacturer on the flow of materials and how frequently shoes are used, upgraded, repaired, and replaced, providing insight into future trends in fashion and style across markets. These shoes are intended to be manufactured as close to the customer as possible through the implementation of decentralized mini factories which would also serve as store fronts and repair centres where worn shoe modules can be replaced, upgraded, or updated. The payment method is flexible and could be subscription based, bundling shoes and services, such as repairs and upgrades, according to the customer's preferences.

1.4.1 SHOE MANUFACTURING AS-IS

In preparation for the Shoe4.0 case study, a generic shoe manufacturing consumer goods As-Is model was created. The following function, Component Manufacturing details the process of making the different parts of the shoe prior to assembly. Shoe components are the output of this function together with production waste. Footwear Assembly and Packaging is combined in the next function, during which shoe components are assembled and packaged in preparation for retail. The finished shoe product is then taken into the Product Sale and Delivery function which substitutes the control of Shelf Life (since shoes don't have a perishable attribute) into a control which, in a sense, represents the fluctuation of supply and availability.

1.4.2 THE SHOE4.0 MODEL

The Shoe4.0 business model starts with the User Profile Creation function. During this function, the customer provides their general information such as input (name, age) and details regarding their preferred payment method. The resources are the tools that the customer will need in order to input their information, mostly a Network

Connection and the Application, which could be in a mobile device app, web app, or in the actual storefront. This function provides the customer information including their subscription and product preferences as output. This output and other information related outputs are centralized in a Data Processing/Analytics function. The main use of the information provided by the initial phase is to activate the Shoe Design function so that the customer can provide personal preferences in the form of customization options for the product. Based on these choices, pricing is determined, and in the background, data is captured and transformed to a format suitable for the 3D printing machine. The resources are in the form of applications to help the customer scan their foot dimensions and additional technology to carry this out if the customer is in store. The main outputs from this function are in the form of shoe specifications and digital documents. Both are, in essence, the same information being transferred in different formats and for different purposes. The shoe specifications provided by the customer are stored in the Data Processing/Analytics function and this same information but in a 3D printer readable format is provided as a digital document to the following function of Shoe Manufacturing, Repairing, and Refurbishing.

The manufacturing function provides the production, repair, and refurbishing actions. Since this is an RdM model, all of these functions can be performed in the same local (in relation to the customer) facility. Sensors and instrumentation are inputs that represent the technology that is to be included in the shoe according to the customer's requirements: the sensors can provide GPS tracking, health monitoring, or others. These are sourced from other manufacturers and thus, the assumption is made that they cannot be produced by the Shoe4.0 manufacturing facility. Raw material is an input, in this case, it is assumed to be the printing material itself since the entire shoe would be manufactured from the least amount of separable materials possible. The 3D printer is the main resource, together with the brand store/manufacturing facility and the service center. As previously mentioned, the service center and the manufacturing facility are concentrated in the same place as the store front. The service center is the place where shoes are refurbished, extra parts are produced, and other services are fulfilled. This same place contains the 3D printing machines that are used for the manufacturing of the shoe. The outputs from this function are the finished shoe, named the intelligent shoe, and shoe parts which may be requested by the customer in order to repair a damaged part of the shoe. Since the shoe is produced in a modular way, different parts can be disassembled for repair. The shoe or shoe parts are then transformed by the Shoe Use function. This function is controlled by delivery or pick-up methods involved in transporting the shoe to the customer and the user type/wearing habits. Additionally, there will be resources provided by the manufacturer in the form of Cleaning, Repair, Re-fashion Services, and a User/Assembly Manual. The possibility of modifications being made by the customer on his own account is aided by the manufacturer-provided use/assemble manual. The outputs provided by this function are digital in the form of wearing data and physical in the form of a worn/damaged shoe. The wearing data is transferred to the Data Processing/Analytics function which uses them as input in order to, for example, activate a service offering or provide other useful information for the manufacturer to support the customer.

The worn/damaged shoe is transferred to the Disposal function. Most of the information generated throughout this process is meant to be what transforms the end of life product into a possible input, as material for the production of other shoes or the end of life product can be recycled/dumped by the customer. The option is free for the customer to choose if they desire to recycle/dump the shoe, but the intention of the Shoe4.0 project is to have the damaged end of life product returned to the manufacturer for reprocessing. For this reason, the brand store/manufacturing facility is included in the resources for this function. Furthermore, there is a consideration for the pick-up/drop-off of the used/damaged shoe/part so that it may reach the place where it will be recycled or reprocessed.

Most of the information generated throughout this process is meant to be capitalized on, to make profit and improve processes, therefore, the Data Processing/ Analytics function is included as part of the model. This function gathers all the information about the customer profile, product specifications, and wearing data by using resources such as a database and datacenter. Through analytics processing of these data sources, the related function can locate the manufacturing facility that is closest to the customer as well as provide contact information.

1.4.3 Validation and Conceptual Model

The Shoe4.0 model was validated through the creation of a questionnaire containing a combination of closed and open questions. The questionnaire was completed by ten experts drawn from both industry and academia. The feedback was requested in order to validate the ability of the Shoe4.0 model to convey the main criteria that was used to create it. For this reason, the validation questionnaire was divided into four sections, each one corresponding to one of the four criteria points of transportation, customer involvement, circularity, and servitisation (identified from the review of relevant literature). The feedback regarding the different sections was used as input for the development of a concept model, called the Shoe4.0 Hybrid Business Model. This model attempts to improve on the Shoe4.0 Model based on the observations gathered from the validation questionnaire and meetings with the Shoe4.0 project members. The main improvements focus around providing a clearer representation of the services and their involvement in the value chain. This was achieved by including the function called Servicing, Re-fashioning, and Repair. This function is controlled by the Data Processing/Analytic function by providing customer details such as their contact information and location, both of which are included due to them being critical for the provision of any service. The output is purely services and service data. The services are now controls, which shape the Shoe Use function by way of providing re-fashioning and/or repair services. The Disposal function is also controlled by services that provide the customer with the option of returning the shoe to the manufacturer once it has reached its end of life. This supports the combination of services with circularity criteria. For this reason, shoe parts from the Shoe Manufacturing, Remanufacturing function and waste material from the Disposal function are inputs to the Servicing function since they are used to provide repairing and take-back services. In addition to the Servicing function, this model has the inclusion of a Component Manufacturing function. This is the reason for which the model is being called a hybrid, since it borrows the function in the shoe manufacturing As-Is, which represents the manufacturing of components that cannot be made, in this case, using additive manufacturing (3D printing).

1.5 DISCUSSIONS AND IMPLICATIONS

The modelling of RdM technologies and methods using IDEF0 has provided insight into the main areas that are influenced by the implementation of such a model in a consumer goods industry. Discussed below is: (1) the reasoning behind the selection of IDEF0 as a tool and its application in modelling a consumer goods manufacturing As-Is value chain, (2) how criteria were developed in order to determine the inputs, outputs, resources, and controls terminology to be used by each model, and (3) what the main insights gained from the comparison between the As-Is and To-Be models are.

Once the model development methodology was defined, a specific modelling technique, in this case IDEF0, was chosen for the following reason: such a technique is used for developing an organizational graphic representation of the business and manufacturing process involved in complex systems such as enterprises. It breaks down the main processes into functions to depict, from a high-level perspective, all the inputs, outputs, controls, and mechanisms or resources that are used by these functions.

The main functions are complemented by four different criteria points which were chosen in order to provide a methodical way of including desired elements for controls and resources which could then be compared between the different models. These criteria are transportation, customer involvement, circularity, and servitisation. These main criteria were chosen because they are the most probable aspects to be affected by the transition from the current way of manufacturing towards a re-distributed model. Therefore, providing knowledge into what are the main aspects that will be influenced in the transition from the current As-Is to a re-distributed To-Be scenario.

1.5.1 TRANSPORTATION CRITERIA

The transportation criteria represent one of the key concepts of RdM. If a manufacturing company wishes to implement a re-distributed model, they are faced with the challenge that the factory and the customer must be located in close vicinity of each other to improve the model's success. Therefore, RdM models with facilities that produce in the proximity of the customer need to be small and flexible.

Transportation can be represented in the models whenever materials are physically moved, providing specific consideration to the distances between where they are procured, manufactured, sold, consumed, and disposed. The aspects considered and included in the models therefore deal with the transportation of raw materials and components from their sources to the places where they become products. The main benefit of creating an As-Is model using these criteria is that the dependence on transportation resources becomes immediately apparent, it is the most prevalent resource affecting almost all functions. Bringing forward this dependency is an important step towards reducing it in the implementation of re-distributed business models.

1.5.2 Customer Involvement Criteria

The second criteria determines which aspects to include in each function in relation to customer involvement. This is a key concept because RdM implies a greater depth of participation by the customer in each of the different stages of a manufacturing value chain. Customer involvement criteria therefore relates to how much participation and decision-making power the customer has in each function. This could be represented by customization and personalization options available to the customer during the different functions, like the option to modify or provide their own designs. These criteria also identify the technology that enables customization and personalization, such as additive manufacturing, design applications, and 3D scanning technology, so they can be included as resources.

With future developments in new manufacturing models and novel materials and technologies, 3D printing can encourage participation by communities in a variety of manufacturing industries and also help motivate innovation. 3D printing technology, given its high degree of digitization and automation, can be integrated into modern models in order to fully realize its benefits. As a technology, additive manufacturing can most certainly lead to the creation of value, but in order to achieve a competitive advantage the current business models need to evolve and allow for the capturing of this value. For this reason, technology such as 3D printing and additive manufacturing must be represented in the models under customer involvement criteria. How much the customer is or is not involved in each function is considered in each model to demonstrate how much the customer defines product and production aspects which can go further than design or aesthetics preferences, such as where and how the product is manufactured, though in the Hybrid model, this customer co-creation is further highlighted due to improvements made in the servitisation aspects. These criteria are crucial to consider when implementing a re-distributed business model. These criteria represent the consumer's ability to participate and influence the value chain process through decision making at an individual basis, namely, the relationship between each customer and the product development process is the main focus of these criteria. These criteria were used in the creation of the models to identify the different functions in the value chain that the customer interacts and the technology which enables this. The relationship between the consumer and the manufacturer and/ or retailer is considered under the servitisation criteria.

1.5.3 Servitisation Criteria

The servitisation criteria analyse not only the service offerings that could be put forward by the manufacturer or retailer to the customer but also express how the relationship between the consumer, manufacturer, and retailer can be deepened and continued after the retail function. The idea of product ownership, product maintenance, and repair are concepts considered by these criteria. These criteria open up new avenues for the company to gather knowledge on the consumer such as through the use of technology embedded in products which can relay information about performance and use.

As opposed to the standard Shoe4.0 model, the Hybrid model shows a clearer representation of the services and their involvement in the value chain. This was achieved by including the function called Servicing, Re-fashioning, and Repair. This function is controlled by the Data Processing/Analytic function by providing customer details such as their contact information and location, both of which are included due to them being critical for the provision of any service. The output is purely services and service data. The services are now controls that shape the Shoe Use function in the way of providing re-fashioning and/or repair services. The Disposal function is also controlled by services that provide the customer with the option of returning the shoe to the manufacturer once it has reached its end of life.

Services permit more interaction and avenues for the company to gather knowledge on their customers while being more sensitive to their needs, effectively building a deeper relationship than if services were not offered. In addition to developing deeper relationships with customers, the information that's gathered from this interaction can drive future product developments, and engineering work. These follow directly from the previously mentioned Customer Involvement criteria since this provides for increased interaction by the customer, but in these criteria, focus is placed on the relationship between the manufacturer and the customer, as opposed to the customer and their potential to influence the products' aspects.

The representation of services and how they improve interaction within the consumer-manufacturer relationship are important criteria to consider when implementing a re-distributed manufacturing business model. This is because value is no longer only associated with the tangible products, customers are increasingly placing value on results, which are in many cases best provided by product-service offerings. From the industrial standpoint, this means that offerings to customers can no longer depend solely on the production of better, faster, and cheaper products. There is now greater demand for "solution systems consisting of product-service components throughout the customer's activity cycle ..." (Vandermerwe, 1990). The modelling of this aspect in IDEF0 provides the visualization of where and how services can have an effect on the creation of positive value for the customer.

1.5.4 CIRCULARITY CRITERIA

The last criteria created to contribute as guides are related to the circular aspects of the business model. The value chain has to become circular, full of feedback in terms of material flows and information to guarantee the sustainability and resilience of the model. Observing the Shoe4.0 case study, circularity can be seen in the feedback of used materials and shoes back into the Shoe Manufacturing, Remanufacturing function. Circularity can also be observed in terms of data being fed back into the system after the user has begun using the product. This valuable data can help improve services and therefore product durability; since it is assumed that the product will enjoy a longer useful life if a suitable maintenance practice is scheduled.

These circularity criteria are paramount to the implementation of re-distributed models due to their transformative effect on value. These criteria therefore pertain mostly to the I/O of each function. Emphasis is made on whether or not this I/O can be reutilized by another function and avoid as much as possible any outputs that exit the model altogether, thus becoming "circular." When designing an IDEF0 representation of the value chain for implementing re-distributed business models, the benefits of circularity can be visualized as materials that would otherwise be considered of no value are repurposed and reinserted into the value chain, namely, outputs can become inputs for other functions. Creating a sustainable, self-sufficient business model.

1.6 CONCLUSIONS

The aim of this chapter was to develop RdM business models using IDEF0 to serve as a guide for the implementation of RdM concepts in the consumer goods industry. To accomplish this goal, it was imperative to understand the current state of consumer goods manufacturing and how elements of servitisation and circularity, together with technology such as additive manufacturing, could potentially influence the organization of manufacturing in terms of location and scale. Criteria were developed around transportation, customer involvement, servitisation, and circularity concepts to support the development of the models and for their comparison. Through the Shoe4.0 case study, information was recollected to create models that depict what an RdM value chain would look like if applied to the shoe manufacturing consumer goods industry. Most importantly, the ways in which the different functions representing supply, production, distribution, and use can communicate with each other through the transformation of materials into products, service offerings, and data, providing a guide for the implementation of re-distributed manufacturing processes into a consumer goods operation, are considered.

The models have shown that there is a need for robust facilities in close proximity to the customer. These facilities are storefronts which can also manufacture, remanufacture, and provide services. This combination can reduce costs, improve sustainability and provide customizable products and services for customers. Furthermore, the reduction in transportation and increase in customer involvement throughout the process are the main elements that would vary the most if a re-distributed model is implemented. A number of study limitations have been identified in the course of this chapter. The implementation of a re-distributed business model would require further study on additive manufacturing to determine what the ideal installed capacity would have to be to satisfy production needs and local demand, considering the lead time and market size. If the aim is to provide the current standard of quantity flexibility according to demand, then this becomes a prerogative. The models developed in this research are limited to providing an initial guide for developing an implementation strategy of re-distributed business. Further study needs to be conducted on the feasibility concerning, for example, costs and technology readiness. The operation of redistribution in practice is based on the current understanding in the research community. Future interpretations of this concept influenced by rapidly changing technology may differ in their implementation.

The hybrid model featured may in future research be further developed into an interactive representation utilising an Agent-Based approach to modelling. Interactive Agent-Based models could allow for real-time data ingest, providing decision makers with an integrated tool on which to evaluate re-distributed manufacturing decisions more rapidly (addressing both financial, logistic and consumer aspects in parallel). As long as the business model is positioned with a high-value proposition, this concept could be profitable. 3D printing could enable high customer engagement resulting in an increased value proposition in return for their willingness to pay extra for a premium product. As seen in the results, the majority of the revenue generated will come from the after-service element of the model, as it is expected that customers will service their trainers after a period of time, whether is a replacement of the sole upper or sensors. Distribution of manufacturing enabled by 3D printing technologies are ideally positioned to support production on-demand as different variations of the same design, production of spare parts, or repair services can be offered (Kohtala, 2015). However, such printing technologies and materials such as SLS printers and Duraform Flex could be very costly for both existent and new market entrants. Future research is needed to improve the technology, reduce the cost of materials, as well as research to decrease 3D printing process time to increase the capacity of the printer, thus reducing the per unit cost of printers. In future work for the authors, the developed business models will be adapted to other industry cases, utilizing the inherent componentization of the process stages achieved through use of the IDEF0 notation.

NOTES

1   www.bmbf.de/upload_filestore/pub/Bufi_2018_Short_Version_eng.pdf
2   www.idef.com/idefo-function_modeling_method/

BIBLIOGRAPHY

Baines, T., Lightfoot, H., Evans, S., & Wilson, H (2007). State-of-the art in product-service systems. *Proceedings of the Institution of Mechanical Engineers, Part B Journal of Engineering Manufacture*, 221, 1543–1552.

Boeer, C., Pedrazzoli, P., Bettoni, A., et al. (2013). *Mass Customization and Sustainability: An Assessment Framework and Industrial Implementation*; Springer: London, UK.

Fazlollahtabar, H. (2016). Parallel autonomous guided vehicle assembly line for a semi-continuous manufacturing system. *Assembly Automation*, 36(3), 262–273.

Fazlollahtabar, H. (2018a). Lagrangian relaxation method for optimizing delay of multiple autonomous guided vehicles. *Transportation Letters*, 10(6), 354–360.

Fazlollahtabar, H. (2018b). Scheduling of multiple autonomous guided vehicles for an assembly line using minimum cost network flow. *Journal of Optimization in Industrial Engineering*, 11(1), 185–193.

Fazlollahtabar, H. (2019a). An effective mathematical programming model for production of automatic robot path planning. *The Open Transportation Journal*, 11–16.

Fazlollahtabar, H. (2019b). Triple state reliability measurement for a complex autonomous robot system based on extended triangular distribution. *Measurement*, 139, 122–126.

Fazlollahtabar, H. (2020). Comparative simulation study for configuring turning point in multiple robot path planning: Robust data envelopment analysis. *Robotica*, 38(5), 925–939.

Fazlollahtabar, H. (2021). Robotic Manufacturing Systems Using Internet of Things: New Era of Facing Pandemics. *Automation, Robotics & Communications for Industry*, 4.0, 82.

Fazlollahtabar, H. (2022). Internet of Things-based SCADA system for configuring/reconfiguring an autonomous assembly process. *Robotica*, 40(3), 672–689.

Fazlollahtabar, H., & Hassanli, S. (2018). Hybrid cost and time path planning for multiple autonomous guided vehicles. *Applied Intelligence*, 48, 482–498.

Fazlollahtabar, H., & Jalali, S.G. (2013). Adapted Markovian model to control reliability assessment in multiple AGV. *Scientia Iranica*, 20(6), 2224–2237.

Fazlollahtabar, H., & Niaki, S.T.A. (2017a). Binary state reliability computation for a complex system based on extended Bernoulli trials: Multiple autonomous robots. *Quality and Reliability Engineering International*, 33(8), 1709–1718.

Fazlollahtabar, H., & Niaki, S.T.A. (2017b). Integration of fault tree analysis, reliability block diagram and hazard decision tree for industrial robot reliability evaluation. *Industrial Robot: An International Journal*, 44(6), 754–764.

Fazlollahtabar, H., & Niaki, S.T.A. (2017c). *Reliability Models of Complex Systems for Robots and Automation*. CRC Press.

Fazlollahtabar, H., & Niaki, S.T.A. (2018a). Cold standby renewal process integrated with environmental factor effects for reliability evaluation of multiple autonomous robot system. *International Journal of Quality & Reliability Management*, 35(10), 2450–2464.

Fazlollahtabar, H., & Niaki, S.T.A. (2018b). Modified branching process for the reliability analysis of complex systems: Multiple-robot systems. *Communications in Statistics-Theory and Methods*, 47(7), 1641–1652.

Fazlollahtabar, H., & Saidi-Mehrabad, M. (2015). Risk assessment for multiple automated guided vehicle manufacturing network. *Robotics and Autonomous Systems*, 74, 175–183.

Fazlollahtabar, H., & Saidi-Mehrabad, M. (2019). *Cost Engineering and Pricing in Autonomous Manufacturing Systems*. Emerald Publishing Limited.

Fazlollahtabar, H., & Shafieian, S.H. (2014). An optimal path in an AGV-based manufacturing system with intelligent agents. *Journal for Manufacturing Science and Production*, 14(2), 87–102.

Fazlollahtabar, H., Mahdavi-Amiri, N., & Muhammadzadeh, A. (2015). A genetic optimization algorithm for nonlinear stochastic programs in an automated manufacturing system. *Journal of Intelligent & Fuzzy Systems*, 28(3), 1461–1475.

Fazlollahtabar, H., Saidi-Mehrabad, M., & Balakrishnan, J. (2015a). Mathematical optimization for earliness/tardiness minimization in a multiple automated guided vehicle manufacturing system via integrated heuristic algorithms. *Robotics and Autonomous Systems*, 72, 131–138.

Fazlollahtabar, H., Saidi-Mehrabad, M., & Balakrishnan, J. (2015b). Integrated Markov-neural reliability computation method: A case for multiple automated guided vehicle system. *Reliability Engineering & System Safety*, 135, 34–44.

Fazlollahtabar, H., Saidi-Mehrabad, M., & Masehian, E. (2015a). Mathematical model for deadlock resolution in multiple AGV scheduling and routing network: A case study. *Industrial Robot: An International Journal*, 42(3), 252–263.

Fazlollahtabar, H., Saidi-Mehrabad, M., & Masehian, E. (2015b). Mathematical model for deadlock resolution in multiple AGV scheduling and routing network: A case study. *Industrial Robot: An International Journal*, 42(3), 252–263.

Fazlollahtabar, H., Saidi-Mehrabad, M., & Masehian, E. (2015c). Mathematical model for deadlock resolution in multiple AGV scheduling and routing network: A case study. *Industrial Robot: An International Journal*, 42(3), 252–263.

Fazlollahtabar, H., Saidi-Mehrabad, M., & Masehian, E. (2021). Robotic industrial automation simulation-optimization for resolving conflict and deadlock. *Assembly Automation*, 41(4), 477–485.

Fazlollahtabar, H., & Saidi-Mehrabad, M. (2015). *Autonomous Guided Vehicles: Methods and Models for Optimal Path Planning*. Germany: Springer.

Ford, S. & Despeisse, M. (2017). Additive manufacturing and sustainability: an exploratory study of the advantages and challenges. *Journal of Cleaner Production*, 137, 1573–1587.

Freeman, R., McMahon, C., & Godfrey, P. (2017). An exploration of the potential for re-distributed manufacturing to contribute to a sustainable, resilient city. *International Journal of Sustainable Engineering*, 10, 260–271.

Geissdoerfer, M., Savaget, P., Bocken, N.M., et al. (2017). The Circular Economy–A new sustainability paradigm. *Journal of Cleaner Production*, 143, 757–768.

Gerring, J. (2016). *Case Study Research: Principles and Practices*; Cambridge University Press: Cambridge, UK.

Ghisellini, P., Cialani, C., & Ulgiati, S. (2016). A review on circular economy: The expected transition to a balanced interplay of environmental and economic systems. *Journal of Cleaner Production*, 114, 11–32.

Hasan, S., Rennie, A., & Hasan, J. (2013). The Business Model for the Functional Rapid Manufacturing Supply Chain. *Studia Commercialia Bratislavensia*, 6, 536–552.

Kastalli, I., & Van Looy, B. (2013). Servitization: Disentangling the impact of service business model innovation on manufacturing firm performance. *Journal of Operations Management*, 31, 169–180.

Kohtala, C. (2015). Addressing sustainability in research on distributed production: An integrated literature review. Journal of Cleaner Production, 106, 654–668.

Lacy, P., & Rutqvist, J. (2015). *Waste to Wealth: The Circular Economy Advantage*; Palgrave Macmillan: New York, NY, USA.

Lieder, M., & Rashid, A. (2016). Review: Towards circular economy implementation: A comprehensive review in context of manufacturing industry. *Journal of Cleaner Production*, 115, 36–51.

MacArthur, E. (2013). *Towards the Circular Economy*; Ellen MacArthur Foundation: Isle of Wight, UK.

Mai, J., Zhang, L., Tao, F., et al. (2016). Customized production based on distributed 3D printing services in cloud manufacturing. *International Journal of Advanced Manufacturing Technology*, 84, 71–83.

Manyika, J., Chui, M., Bisson, P., et al. (2015). *The Internet of Things: Mapping the Value Beyond the Hype*; McKinsey Global Institute: Washington, DC, USA.

Matt, D., Rauch, E., & Dallasega, P. (2015). Trends towards Distributed Manufacturing Systems and Modern Forms for their Design. *Procedia CIRP*, 33, 185–190.

Moreno, M., & Charnley, F. (2016). Can re-distributed manufacturing and digital intelligence enable a regenerative economy? An integrative literature review. *Sustain. Des. Manuf*, 52, 563–575.

Moreno, M., Turner, C., Tiwari, A., et al. (2017). Re-distributed manufacturing to achieve a Circular Economy: A case study utilizing IDEF0 modeling. *Procedia CIRP*, 63, 686–691.

Moreno, M.A., Court, R., Wright, M., et al. (2019). Opportunities for redistributed manufacturing and digital intelligence as enablers of a circular economy. *International Journal of Sustainable Engineering*, 12, 77–94.

Mu, W., Bénaben, F., & Pingaud, H. (2015). A methodology proposal for collaborative business process elaboration using a model-driven approach. Enterprise Information Systems, 9, 349–383.

Pendeville, S., Hartung, G., Purvis, E., et al. (2016). Makespaces: From redistributed manufacturing to a Circular Economy. In *Sustainable Design and Manufacturing, Smart Innovation, Systems and Technologies*; eds: Setchi, R., Howlett, R. J., Liu, Y., & Theobald, P. Springer: Berlin, Germany; pp. 577–588.

Rayna, T. & Striukova, L. (2016). From rapid prototyping to home fabrication: How 3D printing is changing business model innovation. *Technological Forecasting and Social Change*, 102, 214–224.

Shojaeifar, A., Fazlollahtabar, H., & Mahdavi, I. (2016). Decomposition versus Minimal Path and Cuts Methods for Reliability Evaluation of an Advanced Robotic Production System. *Journal of Automation Mobile Robotics and Intelligent Systems*, 10(3), 52–57.

Stahel, W.R. (2016). The circular economy. *Nature News*, 531, 435.

Sugiyama, H. & Hirao, M. (2014). Integration framework for improving quality and cost-effectiveness in pharmaceutical production processes. In In. J. J. Klemeš, P. S. Varbanov & P. Y. Liew (Eds.): Proceedings of the 24th European Symposium on Computer Aided Process Engineering, Elsevier, *Computer Aided Chemical Engineering*, 33, pp. 379–384 (2014) ; Elsevier: Amsterdam, The Netherlands.

Vandermerwe, S. (1990). The market power is in the services: Because the value is in the results. *European Management Journal*, 8, 464–473.

Yin, R.K. (2013). *Case Study Research: Design and Methods*, 4th ed.; SAGE: Los Angeles, CA, USA.

Zott, C., Amit, R., & Massa, L. (2011). The business model: Recent developments and future research. *Journal of Management*, 37, 1019–1042.

2 Automated Production Systems

2.1 INTRODUCTION

The analysis of safety issues associated with the transportation of hazardous materials is considered to be as important as the study of risk problems connected with fixed installations, since historical evidence has shown that incidents related to the transportation of dangerous goods are comparable in number and magnitude to those that have occurred in chemical plants. Furthermore, hazardous materials releases during transport may occur in areas that are not sufficiently controlled or protected, such as zones of high population density or of natural and historical beauty (CCPS, 1995). For this reason risk analysis should address both fixed installations and transportation networks, thus obtaining a complete area risk evaluation, which represents the basis on which decision makers will establish criteria for risk management, risk control and planning.

AGVs are widely used in industries such as the car industry and other assembly lines. Often those AGVs are rail mounted meaning they are moving along a fixed path. In contrast to the fixed path AGVs, the free path AGVs are required to move around and navigate on their own. Such AGVs are being used in the port of Rotterdam. They have several complex embedded systems in order to fulfill the requirements and meet the expected benefits. For instance, there is an onboard controller that is responsible for propulsion, braking, etc. Further, a management system is also included, which handles traffic control, planning, scheduling, dispatching and routing. In addition to that, the AGVs have a communication system, which is used to send and receive data to and from a central controller. The information that is sent is current position, heading and status of the vehicle. Finally, there is the navigation system that consists of several different scanners thus allowing the AGVs to navigate throughout the terminal to make pickups/drop offs and to avoid collisions and deadlocks.

Flexible Manufacturing Systems (FMSs), container terminals, warehousing systems, and service industries, including hospital transportation, are employing autonomous guided vehicles (AGVs) for material handling to maintain flexibility and efficiency of production and distribution. For efficient operation, it is necessary to realize the synchronized operations for the simultaneous scheduling of production systems and transportation systems (Nishi et al., 2011).

DOI: 10.1201/9781003400585-2

Path planning under uncertainty has been extensively studied in the areas of computer science and operation research due to its wide applications such as communication routing and transportation engineering (see, e.g., Kouvelis and Yu, 1997; Waller and Ziliaskopoulos, 2002; Montemanni et al., 2004). The omnipresent uncertainty of travel times on network links could result from network congestions, hardware failures or traffic jams, temporary construction projects, weather conditions, traffic accidents, and the like. Because of the lack of updated on-line information on network links (cables, roads, etc.), the client/traveler will have to consider possible ranges of the travel times on network links when making the appropriate route choice, that often involves tradeoffs between travel times and risk-sums to be taken (Chen et al., 2009).

Risk-based route evaluation is necessary for decision making by authorities for regulating hazardous materials transport vehicles. The paper highlighted risk estimation and its representation for three highway study routes in western India using frequency analysis through logic diagrams and scenario-based detailed consequence analysis of accidental release categories (Chakrabarti and Parikh, 2013).

Shipments of hazardous materials underline the significant risk and tremendous cost, but literature has focused only on the cost-effective scheduling of these vehicles. It is important that transport companies consider risk since the insurance premiums are contingent on the expected claim. Siddiqui and Verma (2015) presented a mixed-integer optimization program with operating cost and transport risk objectives, which could be used to prepare routes and schedules for a heterogeneous fleet of vehicles. The bi-objective model was tested on a number of problem instances of realistic size, which were further analyzed to conclude that the cheapest route may not necessarily yield the lowest insurance premiums, and that larger vessels should be used if risk is more important as it enables better exploitation of the risk structure.

Effective design of automatic material handling devices is one of the most important decisions in flexible cellular manufacturing systems. Minimization of material handling operations could lead to optimization of overall operational costs. An AGV is a driverless vehicle used for the transportation of materials within a production plant partitioned into cells. The tandem layout is according to dividing workstations into non-overlapping closed zones to which a Tandem Automated Guided Vehicle (TAGV) is allocated for internal transfers. Shirazi et al. (2010) illustrated a non-linear multi-objective problem for minimizing the material flow intra and inter-loops and minimization of maximum amount of inter cell flow, considering the limitation of TAGV work-loading. For reducing variability of material flow and establishing balanced zone layout, some new constraints have been added to the problem based on the six sigma approach. Due to the complexity of the machine grouping control problem, a modified ant colony optimization algorithm was used for solving this model.

Fazlollahtabar and Shafieian (2014) concerned with the design of a computer integrated manufacturing system to identify an optimal path in a Vehicle Routing Problem (VRP) network addressing triple criteria. In most VRPs just one criterion, either time or cost was considered in decision making. They considered all time, cost, and AGV capability in decision making, simultaneously. To satisfy Material Requirement Planning (MRP) by providing the Bill of Material (BOM), AGVs are suitable devices.

Fazlollahtabar et al. (2015) were interested in both time and cost optimization in an automated manufacturing system. In their proposed system, material handling is carried out by AGVs and robots for the manufacturing functions. In this way, defect rates, breakdown times, waiting time, processing time, and certain other parameters were not expected to be deterministic. Stochastic programming was applied to optimize the production time and material handling cost. The proposed stochastic program was a nonlinear model, for this reason a Successive Linear Programming (SLP) technique was employed for its optimization. Numerical test results pointed to the inefficacy of the proposed optimization method for large sized problems. Hence, a Genetic Algorithm (GA) was presented to optimize large sized problems.

The optimization of Material Handling Systems (MHSs) can lead to substantial cost reductions in manufacturing systems. Choosing adequate and relevant performance measures is critical in accurately evaluating MHSs. The majority of performance measures used in MHSs are time-based. However, moving materials within a manufacturing system utilize time and cost. Tavana et al. (2014) considered both time and cost measures in an optimization model used to evaluate an MHS with AGVs. They took into account the reliability of the MHSs because of the need for steadiness and stability in the automated manufacturing systems. Reliability was included in the model as a cost function.

One of the principal matters of concern in the multiple AGV-based industrial system is the prevention of conflicts between AGVs which might escalate to collision. Although AGV collisions have actually been very expensive events, contributing to a very small proportion of the total fatalities, they have always caused relatively strong impacts mainly due to a relatively large number of fatalities per single event and occasionally the complete failure of the industrial system.

The main driving force for developing risk methods/models during the 1960s was the need for increasing the efficiency of the system. In general, separating AGVs using space and time separation standards has prevented conflicts and collisions. However, due to the reduction of this separation, in order to increase the capacity of AGVs and thus cope with growing industrial transport demand, assessment of the risk of conflicts and collisions under such conditions has been investigated using several important methods/models. The methods/models were expected to show whether a reduction of separation and spacing between AGVs' handling task tracks would be sufficiently safe, i.e., determine the appropriate spacing between tracks guaranteeing a given level of safety.

The main purpose of collision risk analysis has always remained as being to support decision-making processes during system planning and development, through evaluation of the risk and safety of the proposed changes (either in the existing or the new system).

2.2 AUTOMATION EFFICIENCY

Sustainability is a term that now permeates, with all its nuances, our workplaces, homes, schools and many other organized groups, as a true pop star, bringing a sense of responsibility and a utopian, or not, search for a global society, conscious of its actions and interdependence. A term so common today, but which was despised some years ago, being relegated to the worst times of Brazilian television, for it to keep something in its program grid, focused on sustainable development. The evolution of the sustainability concept, particularly, draws attention and is needed to guide and motivate those who either directly or indirectly are involved in self-sustainable projects, or are opinion leaders. However, after all, how is sustainability defined? A quick search in the dictionary does not help much. This may lead us to reflect on the most basic concepts such as the maintenance of environmental balance or for social responsibility, amongst other ideas. All these and many other ideas are partially correct. We will look at how these various issues have culminated in the concept of sustainable development and how automation and control play a key role in the implementation and maintenance of sustainability. These disparate ideas and studies began to gain momentum and political force in the late 60's, with the creation of the Club of Rome. This group, formed by key people in various sectors and countries, began sponsoring studies and advanced research related to profitable and sustainable economic growth.

AGV technology has become an interesting study area in order to increase the performance and utilization of the container terminal. In this paper, we are primarily focused on the dispatching problem because of its significant importance when it comes to performance in a container terminal with an AGV system. Previous research regarding dispatching AGVs has not considered the usage of cassettes, and to the best of our knowledge, no study has yet been conducted regarding the dispatching problem for the AGV system.

One of the most distinct differences between using cassettes together with AGVs is that the cassettes have the advantage of working as a buffer at both the pick-up and the delivery point. One of the benefits with this feature is that a container can be loaded or unloaded without the presence of an AGV. This feature allows more flexibility into the dispatching strategies due to the fact that the cranes working in the container terminal do not have to wait to be served by an AGV.

Hence, this study's primary goal is to consider already existing dispatching strategies for other AGV systems, in order to construct dispatching strategies that can be used together with this newly developed AGV system. During the creation of the simulation model, the following aspects have been considered:

- Flow path layout
- Traffic management: prediction and avoidance of collisions and deadlocks
- Number and location of pick-up and delivery points
- Vehicle requirements
- Vehicle dispatching
- Vehicle routing.

The first result of this initiative was the publication of a report commissioned in 1972 by MIT researchers called, "The Limits to Growth", which dealt with crucial points for the development of humanity through energy, pollution, sanitation, health, environment, technology and population growth. Through mathematical modeling, these studies revealed that by 2100, the Earth will not support the uncontrolled population growth due to impacts on the above points.

From this mutual understanding between developed and underdeveloped nations in relation to industrial activities, the actions that followed were supported by organizations such as UNESCO, which created in 1975 the International Program on Environmental Education (IIEP) focused on training educators to follow the principles of continuous environmental education, adapted to regional differences and aimed at fulfilling the interests of the Nations. In 1980, the International Union for Conservation of Nature published the report "The Global Strategy for Conservation," where the concept of "sustainable development" appeared for the first time. Up to this point, it seemed that everything was well, but in 1987, the Brundtland report (or "Our Common Future) pointed out the incompatibility between the models of sustainable development and the current means of production and consumption. Then through a series of new measures, it was finally possible to formalize the concept of sustainable development as "the development that meets present needs without compromising the ability of future generations to meet their own needs".

Among the various measures indicated below, only a few are already a reality in our daily lives:

- Limitation of population growth
- Guarantee of basic resources (water, food, energy) in the long term
- Preservation of biodiversity and ecosystems
- Reduction of energy consumption and the development of technologies using renewable energy sources
- Increasing of the industrial production in non-industrialized countries based on environmentally adapted technologies
- Control of unplanned urbanization and integration between countryside and smaller towns
- Meeting of the basic needs (health, education, and housing)
- Adoption of the sustainable development strategy by development organizations (agencies and international financial institutions)
- Protection of supranational ecosystems such as Antarctica, the oceans, etc., by the international community
- Outlawing of wars
- Implementation of a sustainable development program by the United Nations (UN).

Based on these corrected guidelines, the concept of sustainable development should be adopted by business leaders as a new way of producing without degrading the environment. Once assimilated, this culture should then be disseminated among the organizations, extensible to all levels, in order to identify the impact of its production on the environment, resulting in the execution of a project that combines production and environmental preservation with the use of new technologies.

Some other measures for the implementation of a minimally adequate program for sustainable development are:

- Use of new construction materials
- Restructuring of residential and industrial areas

- Use and consumption of alternative energy sources such as solar, wind and geothermal
- Recycling of reusable materials
- Rational consumption of water and food
- Reduced use of chemical products harmful to health in food production.

With clearer and more coherent goals, the conferences began having greater impact and coverage on their agendas, a fact confirmed in 1992 at the UN Conference on Environment and Development, the birthplace of Agenda 21, which approved the Convention on Climate Change, the Convention on Biological Diversity (Rio Declaration - ECO-92), as well as the Statement of Principles for the Sustainable Management of Forests.

To compensate for and make the efforts more transparent and valuable, the Dow Jones Sustainability World Index was created in 1999, being the first indicator of the financial performance of leading companies in global sustainability. Their derivations now include companies in the stock exchange of the United States and Europe. Following the same methodology, in 2005 the Corporate Sustainability Index (ISE) was created, as the first initiative in Latin America (Bovespa), one that represented an efficient way for the economic sector to encourage sustainability initiatives in Brazilian companies. The term "sustainability" now has a wider connotation and is spreading rapidly, now incorporated in the politically correct vocabulary of business, mass media, civil society organizations, almost achieving a global unanimity. However, the solution to the causes for unsustainability seems to move at a much slower pace, even with more dire predictions about the future. This apparent slowness occurs even with the encouragement of debates, whose proposed solutions may conflict with the various parties involved.

Otherwise, within the social perspective, we face our own challenges. Just to give you an idea, according to recent surveys in Brazil, from harvesting to community consumption, 20% of all food produced in the country is wasted (IBGE statistical agency). That would be enough to supply all of the Brazilian poor. Moreover, they generate 125 000 tonnes of organic waste and recyclable materials a day. Not to mention the waste of 50% of treated water throughout the country. Likewise 9.5% of the annual energy production is also wasted. These are impressive numbers. The Brazilian Association of Maintenance indicates that maintenance costs in the country reach the milestone of 4.2% of GDP, and 4% of gross sales of companies are spent on maintenance activities. These data are more than sufficient for this complicated scenario, full of challenges, where the engineering of automation can provide the necessary support to these and many other demands of society, always looking for the "optimal" balance point of this equation whose three main variables are the ecological, social and economic requirements. The figure below shows deviations of results, when one of the main variables of this "equation" is put aside. The results obtained are very different from what sustainability means:

2.3 MANUFACTURING AUTOMATION EVOLUTION

Manufacturing automation, often referred to as factory automation or industrial automation uses a combination of software, computers, and robotics (or control systems) in place of human labour. Economic and global shifts affect manufacturing trends and the changing needs and evolution of the industry are matched by changes in manufacturing automation. As the world becomes more acutely aware of the effects of climate change, demands for more sustainable manufacturing practices are on the rise. Manufacturing automation companies have helped the industry at large keep pace with this demand. This can be seen through the development and use of Robotic Process Automation (RPA) to achieve 'Smart Manufacturing' practices. Smart Manufacturing connects various control systems through the Internet of Things (IoT). Using a combination of data analytics and machine learning, these devices work together to create a more efficient and sustainable process.

Companies that use manufacturing automation experience increases in accuracy, flexibility, and productivity with decreases in energy consumption, waste, and workplace injury. These benefits mean that manufacturing automation can bolster a company's bottom line and provide a more sustainable approach to manufacturing. Although constructing a sustainable factory may require more upfront capital, the reduced operating expenses, increased productivity, and increased flexibility made available to companies, through 'Smart Manufacturing' means long-term savings. Many manufacturers from breweries to pulp and paper mills are catching on to the environmental benefits of manufacturing automation.

Advances in automation over recent years have given rise to the promise and potential of Autonomous Production. While the vision of 'thinking' production lines has still to be realized, Industry 4.0 enabling technologies, such as such as the IoT (Internet of Things), Machine Learning and Cyber Physical Systems (CPS) in industrial settings present concrete opportunities towards more responsive, smarter and more efficient production. The route to autonomous production requires the integration of complex systems and the collection and analysis of multiple data streams. At the same time, increasing pressure on the environment from human activity necessitates a shift in attitudes to the way industry currently operates. The rise of the circular economy is one specific response to this need, promoting a holistic view of both production and consumption of goods and services. Automation and autonomous manufacturing provides a new way of looking at production where data evidence may be analysed and acted upon in real time leading to the potential for reductions in waste, longer reliable usage patterns for products, predictive monitoring of industrial processes and whole life consideration of products with particular regard to the recycling of defunct products and their remanufacture.

The use of machine learning enables insights from the production line to be systematically captured and employed by human experts for decision-making. This is a crucial step in the development of fully autonomous production lines and promotes methods to identify more efficient and environmentally acceptable processes derived from sensed data collected from both inside and outside the organization along with data mined from existing data stores. With intelligent systems use comes the need to explain the reasoning behind the results they produce to humans for the purposes of decision support, ensuring provenance and maintaining quality control. This need for explainable AI (Artificial Intelligence) in manufacturing systems supports the concept of the 'human in the loop' to enable a new level of informed decision making to take place.

2.3.1   Flow Path Layout

The flow path aspect refers to the design of the paths that the AGVs can travel on. The layout of a flow path connects the different locations that can be found in container terminals such as machines, processing centers, stations and other fixed structures along the aisles. Designing the layout of a flow path can be done in various ways with different considerations. However, a layout is usually represented by a directed network where the nodes are considered as aisle intersections, pick-up and delivery locations. Arcs in the layout represent the guided paths on which the AGVs can travel. These arcs determine the direction of travel, which can be unidirectional or bidirectional. Unidirectional arcs are just allowing the vehicles to travel in one direction, and bidirectional arcs allow vehicles to travel in both directions. Using bidirectional arcs can obtain a reduction in travel distance, due to the fact that AGVs are able to make shortcuts. However, unidirectional paths are much easier to construct and control. Another solution is to use multiple lanes where opposite direction can be inserted in the two arcs. The downside of using multiple lanes is that it demands more space, which may not be available.

2.3.2   Traffic Management

Traffic management addresses the problem of preventing AGV collisions and deadlocks. It is crucial that an AGV has the ability to return to its original path after avoiding an obstacle without causing any collisions. By placing sensors on the AGV, it is able to detect obstacles, which prevent collisions. Deadlocks can occur in systems that uses bidirectional flow paths, for example, when two AGVs are forced to stop because they are moving in opposite directions on the same path. In addition to deadlocks in bidirectional flow paths, they can also occur at the area of pick-up and delivery. Prevention of collisions and deadlocks can be solved by rerouting of vehicles. However, the performance of the system will decrease with high occurrences of rerouting, and therefore preplanning the avoidance of deadlocks and collisions is a much better solution. Previous studies have resulted in methods to avoid these aspects. If we look at the last category of routing strategies, two distinct algorithms can be developed for this purpose, namely static and dynamic. The different routes in static algorithms are predetermined, for example, from x to y, the shortest path will always be used. Dynamic routing algorithms will on the other hand also consider traffic conditions and adapt to changes within the system before determining the most profitable path. Much of the previous work in this area has used the well-known Dijkstra's shortest path algorithm incorporated with bespoke strategies to handle routing. However, areas such as distribution, transshipment and transportation systems have hardly been studied in the context of routing AGVs.

2.3.3   Pickup and Delivery Points

Another aspect is the location of pickup and delivery points, which mainly refers to the locations of cranes, machines, inspection stations and places of storage. These locations for an AGV system will influence the performance of the system, for example the distance that the AGVs need to travel. This aspect is even more important for large AGV systems to avoid bottlenecks between pick-up and delivery. However, this area is still somewhat unexplored, and further studies need to be conducted in order to decide whether previously developed strategies are applicable for large AGV systems.

2.3.4 Vehicle Requirements

The fourth aspect regarding AGV systems is the vehicle itself, and its requirements. The vehicle's characteristics such as speed, capacity and cost must be considered along with other factors such as the flow path layout, routing and dispatching strategies in order to determine the number of vehicles required for that specific system to obtain the most optimized system. Looking more closely at the function of controlling the AGV, there are mainly two aspects linked to that, namely routing and dispatching. Both of these aspects are highly important for any control policy within any AGV system, in order to obtain the objectives of satisfying the demand for the fastest transportation possible without the occurrence of conflicts between AGVs.

2.3.5 Dispatching and Selection of Jobs

As mentioned previously in this chapter, routing refers to traffic management and how AGVs are routed between the different locations within the system without deadlocks and collisions. The dispatching concept refers to the selection and assignment of tasks to the AGVs. In general, the main goal of the selection of a job is to minimize the cost, which can be a combined value of time, distance and priority. To determine which AGV is suited for a specific job, various methods can be used in various ways. In the next chapter, we will review and present previously studied methods that can be used for dispatching and job selection regarding AGV systems in a container terminal. However, these methods have been constructed for AGV systems without the consideration of cassettes.

2.4 STATEMENT OF THE PROBLEM

In an industrial system having multiple AGVs as material handling devices, a significant problem is path planning so that collision among AGVs is minimized. For financial analysis it is more effective to obtain an estimation of the losses before or during implementation of such a system. A very helpful method to handle uncertainty is risk analysis. Consider an industrial manufacturing system having several AGVs as material handling tools. AGVs are dispatched according to a request from a station. Products are processed in stations according to process plan and job sequences. Multiple stations may request AGVs at the same time and thus a precise path plan is essential to avoid collisions of AGVs. Also, AGV's arrival within a set timeframe at a station satisfies just in time manufacturing and therefore avoiding delays is important. AGVs on some paths may become obstacles for others. Consequently, a complicated system of AGVs, paths, and stations is configured leading to possible collisions and obstacles. The problem is difficult for hazardous manufacturing systems (e.g.

galvanization or plating production systems). The aim in these cases is to provide separate path plans for each AGV with minimal risks. The risk is associated with collision and obstacle avoidance of AGVs. The risk is defined as the product of probability of collision and the severity of a collision if it takes place.

2.4.1 SIGNIFICANCE OF THE MODEL

One of the extensive applications of industrial robots is in hazardous environments. Designing an appropriate control mechanism to fulfil handling tasks efficiently is significant. Therefore, the proposed problems and models are adaptive to process AGVs in such environments.

2.4.2 MAJOR CONTRIBUTION

As reviewed in the literature, most of the AGV path planning activities considered either time or cost and in some cases both time and cost. A research gap is noticeable on how to include quality in the path plan which is a critical performance criterion. Then, time, cost and quality simultaneously are modelled and optimized. Quality is considered using risk. Also, considering multiple AGVs is another aspect of novelty of this chapter.

2.5 MODELING THE PROBLEM

Modelling the problem into the mathematical modelling space is needed for solving the collision avoidance problem through risk modelling and optimization. In this problem, an AGV moves from one specific point to another specific point, from start to end points, through a particular route where there is some obstacle in its route. The AGV is modeled according to its central coordinates and as a point that moves from start to end through a specific set of coordinates. The AGV has constant speed and constant direction for a move between the two points. The static obstacles are modeled as a specific range of coordinates. If the movement coordinates of the AGV is the same as the static obstacle coordinates, at least at one point, then a collision happens. The dynamic obstacles are modeled as moving points with constant speed and constant direction. The coordinates of the dynamic obstacles will be calculated dynamically according to the AGV's position. In order to avoid a collision the AGV will change the speed and/or the direction of its movement at one point that is referred to as a redirection point. For presenting and modeling, the route of the AGV can use the coordinates of redirection points along with the speed and direction of the AGV's movement.

The mathematical formulations are presented below.

Indices:

i, j	Counter for stations, $i = 1, \ldots, n; j = i + 1, \ldots, n$
k, k'	Counter for AGVs, $k, k' = 1, \ldots, m$

Parameters:

s_{ijk}^c Severity of collision hazard of AGV k in path i to j

p_{ijk}^c Probability of collision of AGV k in path i to j

R_{ijk}^c Risk of collision of AGV k in path i to j

t_{ij} Required time to move in path i to j

d_{ij} The distance between station i to station j

d_{0k} Current position of AGV k

v_k The velocity of AGV k

a_k The acceleration of AGV k

Decision Variables:

$$x_{ijk} = \begin{cases} 1 & \text{If AGV } k \text{ is assigned to path } i \text{ to } j \\ 0 & \text{Otherwise} \end{cases}$$

$$y_{kk'} = \begin{cases} 1 & \text{If AGV } k \text{ collide with AGV } k' \\ 0 & \text{Otherwise} \end{cases}$$

$$z_{jk} = \begin{cases} 1 & \text{If station } j \text{ requests AGV } k \\ 0 & \text{Otherwise} \end{cases}$$

$$\min W = \sum_{k,k'=1}^{m} \sum_{j=i+1}^{n} \sum_{i=1}^{n} R_{ijk}^c \cdot y_{kk'}. \tag{1}$$

Such that:

$$R_{ijk}^c = p_{ijk}^c \times s_{ijk}^c, \tag{2}$$

$$d_{ij} = \frac{1}{2} a_k . t_{ij}^2 + v_k . t_{ij} + d_{0k}, \tag{3}$$

$$\sum_{i=1}^{n} x_{ijk} = z_{jk}, \quad \forall k, j, \tag{4}$$

$$\sum_{j=i+1}^{n} \sum_{i=1}^{n} x_{ijk} = 1, \quad \forall k, \tag{5}$$

$$\sum_{k=1}^{m} z_{jk} = 1, \quad \forall j, \tag{6}$$

$$\sum_{k=1}^{m} x_{ijk} d_{ij} \leq \min_{k} \left\{ d_{0k} z_{jk} \right\}, \quad \forall i, j, \tag{7}$$

$$\sum_{k=1}^{m} d_{0k} z_{jk} \leq \sum_{k'=1}^{m} d_{0k'} z_{jk'}, \quad \forall j, \tag{8}$$

$$\sum_{k=1}^{m} p_{ijk}^{c} \leq \sum_{k,k'=1}^{m} \frac{y_{kk'}}{x_{ijk}}, \quad \forall i, j, \tag{9}$$

$$y_{kk'} \leq \sum_{j=i+1}^{n} \sum_{i=1}^{n} x_{ijk} z_{jk}, \quad \forall k, k', \tag{10}$$

$$\sum_{k'=k+1}^{m} \sum_{k=1}^{m} y_{kk'} . a_k \leq \sum_{k=1}^{m} s_{ijk}^{c}, \quad \forall i, j, \tag{11}$$

$$\sum_{k'=k+1}^{m} \sum_{k=1}^{m} y_{kk'} < \sum_{k=1}^{m} z_{jk}, \quad \forall j, \tag{12}$$

$$x_{ijk}, z_{jk}, y_{kk'} \in \{0,1\}, \quad \forall i, j, k, k'. \tag{13}$$

Equation (1) is the objective function of the problem minimizing the risk of collisions. Relation (2) computes the risk including the probability and the severity of a collision hazard. Relation (3) shows the distance computation using the acceleration and velocity of AGVs. Equation (4) guarantees the simultaneous assignments of AGVs to stations and requests from stations. Equation (5) assures that an AGV is assigned to a station. Equation (6) emphasizes that for a station one AGV is requested. Equation (7) shows that the request rule is based on the closest distance of an AGV to a station. Equation (8) shows the sequence of AGVs requested to stations. Equation (9) computes the probability of collision. Equation (10) guarantees that if an AGV is assigned to a station based on a request then the collision is avoided correspondingly. Equation (11) shows the computation of the severity of collision hazard based on the acceleration of the AGV's movement. Equation (12) specifies that if an AGV is requested by a station then it does not collide with others. Relation (13) shows that the decision variables are binary.

2.6   AUTOMATION GOES GREEN IN INDUSTRIAL NETWORKS

Only the balance of these three variables produces a self-sustaining solution. In addition, the process automation is one of the enablers for a sustainable reality. Only the balance of these three variables produces a self-sustaining solution. Moreover, the process automation is one of the enablers for a sustainable reality. The technological support of global organizations in the standardization of equipment resources is one example. Today, entire equipment and automation solutions leave the factory with the green seal of sustainability, not only because they consume less energy but also because they have been designed with a standard application profile that allows the plant to turn itself off in unused areas. An example of this concept is the ProfiEnergy, the newest Profibus International application profile. It covers not only the basic procedures, such as control loop fine tuning, but also other aspects to improve processes such as asset management systems, Manufacturing Execution Systems (MES) and Business Intelligence, whose objectives are clearly outlined: increasing production capacity without investing in expansions and/or new plants, bringing operational availability to its peak, avoiding unplanned downtime, wastefulness with final product variability and failures in control production (due to lack of real-time information for decision makers). With industrial networks and intelligent equipment, the basis of industrial automation now has data that previously would be topics for science fiction movies. Non-Control information is increasingly available and ready to meet these specialized systems.

Previously known as a Manufacturing Execution System (MES) it is now officially the Manufacturing Enterprise Solution (MES) for representing much more than just a system for production control. Issues such as product quality, inventory, maintenance, data management and life cycle management cannot be analyzed separately from the manufacturing control.

2.6.1 ENVIRONMENTAL BENEFITS OF MANUFACTURING AUTOMATION

Automated manufacturing is inherently sustainable. Its existence allows companies to create and use processes that minimize the negative environmental impact. This helps companies to achieve sustainable manufacturing goals including lowered energy, water, and land usage.

Here are some of the ways that manufacturing automation companies help the environment:

- Data collected is used by manufacturers to ensure their machines are efficient.
- More efficient machines have lower heating requirements.
- Robots help to streamline processes, meaning less material waste.
- Automation allows for reduced cycle times, resulting in reduced energy output.
- Robots can be compact and can streamline processes thus requiring less factory floor space.

From a business perspective, manufacturing automation can help companies reach Key Performance Indicators (KPIs) related to corporate sustainability and environmental regulatory compliance. A combination of regulations and internal guidelines has resulted in industrial companies seeing cost savings and increased efficiency.

Sustainable manufacturing and automation are two pieces of a very large environmental puzzle. Companies who grow and evolve with an eye to the future will see that investing in sustainable innovations is necessary to become economically competitive. This leads to new job opportunities and a more stable future for both the company and the environment. Adverts from manufacturing automation companies such as Innovative Automation might say: We support sustainable manufacturing through automation and have the experience and skills necessary to tackle automation projects. We are experts in a wide variety of manufacturing automation applications including assembly automation, testing automation, robotics and welding. Contact us today to discuss how a sustainable manufacturing automation solution can help your business.

2.7 CONCLUSIONS

Sustainability goes far beyond Advertising and Publicity. It's a simple concept to understand, but it brings an entire historical context of conflicting interests and the pursuit of understanding between the involved parties. In this search, much has been learned and much will be learned. The speed of adoption of sustainable solutions is well below the desired level, as it requires political will to force changes; business will to create significant improvements in their means of production in addition to the determined pressure by society as a whole. Society, by the same token, must adopt an awareness of sustainability and avoid uncontrolled consumerism. You cannot, however, belittle the achievements gained so far, as it is a very difficult concept to keep in the absence of tools and policies. In this context, automation plays an important role. Through further research and the adoption of smarter methods and production tools this situation tends to improve very much, with little investment. Finally, we are all responsible for reinventing our future in a self-sustaining way, focusing on the common well-being of our planet.

BIBLIOGRAPHY

CCPS, (1995). *Guidelines for Chemical Transportation Risk Analysis*, AIChE, New York.

Chakrabarti, U.K., & Parikh, J.K. (2013). Risk-based route evaluation against country-specific criteria of risk tolerability for hazmat transportation through Indian State Highways. *Journal of Loss Prevention in the Process Industries* 26, 723–736.

Chen, X., Hu, J., & Hu, X. (2009). A new model for path planning with interval data. *Computers & Operations Research*, 36, 1893–1899.

Fazlollahtabar, H. (2016). Parallel autonomous guided vehicle assembly line for a semi-continuous manufacturing system. *Assembly Automation*, 36(3), 262–273.

Fazlollahtabar, H. (2018a). Lagrangian relaxation method for optimizing delay of multiple autonomous guided vehicles. *Transportation Letters*, 10(6), 354–360.

Fazlollahtabar, H. (2018b). Scheduling of multiple autonomous guided vehicles for an assembly line using minimum cost network flow. *Journal of Optimization in Industrial Engineering*, 11(1), 185–193.

Fazlollahtabar, H. (2019a). An effective mathematical programming model for production of automatic robot path planning. *The Open Transportation Journal*, 13(1), 11–16.

Fazlollahtabar, H. (2019b). Triple state reliability measurement for a complex autonomous robot system based on extended triangular distribution. *Measurement*, 139, 122–126.

Fazlollahtabar, H. (2020). Comparative simulation study for configuring turning point in multiple robot path planning: Robust data envelopment analysis. *Robotica*, 38(5), 925–939.

Fazlollahtabar, H. (2021). Robotic manufacturing systems using Internet of Things: New Era of facing pandemics. *Automation, Robotics & Communications for Industry*, 4.0, 82.

Fazlollahtabar, H. (2022). Internet of Things-based SCADA system for configuring/reconfiguring an autonomous assembly process. *Robotica*, 40(3), 672–689.

Fazlollahtabar, H., & Hassanli, S. (2018). Hybrid cost and time path planning for multiple autonomous guided vehicles. *Applied Intelligence*, 48, 482–498.

Fazlollahtabar, H., & Jalali, S.G. (2013). Adapted Markovian model to control reliability assessment in multiple AGV. *Scientia Iranica*, 20(6), 2224–2237.

Fazlollahtabar, H., & Niaki, S.T.A. (2017a). Binary state reliability computation for a complex system based on extended Bernoulli trials: Multiple autonomous robots. *Quality and Reliability Engineering International*, 33(8), 1709–1718.

Fazlollahtabar, H., & Niaki, S.T.A. (2017b). Integration of fault tree analysis, reliability block diagram and hazard decision tree for industrial robot reliability evaluation. *Industrial Robot: An International Journal*, 44(6), 754–764.

Fazlollahtabar, H., & Niaki, S.T.A. (2017c). *Reliability Models of Complex Systems for Robots and Automation*. CRC Press.

Fazlollahtabar, H., & Niaki, S.T.A. (2018a). Cold standby renewal process integrated with environmental factor effects for reliability evaluation of multiple autonomous robot system. *International Journal of Quality & Reliability Management*, 35(10), 2450–2464.

Fazlollahtabar, H., & Niaki, S.T.A. (2018b). Modified branching process for the reliability analysis of complex systems: Multiple-robot systems. *Communications in Statistics-Theory and Methods*, 47(7), 1641–1652.

Fazlollahtabar, H., & Olya, M.H. (2013). A cross-entropy heuristic statistical modeling for determining total stochastic material handling time. *The International Journal of Advanced Manufacturing Technology*, 67, 1631–1641.

Fazlollahtabar, H., & Saidi-Mehrabad, M. (2015a). Risk assessment for multiple automated guided vehicle manufacturing network. *Robotics and Autonomous Systems*, 74, 175–183.

Fazlollahtabar, H., & Saidi-Mehrabad, M. (2015b). Methodologies to optimize automated guided vehicle scheduling and routing problems: A review study. *Journal of Intelligent and Robotic Systems*, 77, 525–545.

Fazlollahtabar, H., & Saidi-Mehrabad, M. (2015c). *Autonomous Guided Vehicles: Methods and Models for Optimal Path Planning*. Springer International Publishing, Switzerland. ISBN 978-3-319-14746-8.

Fazlollahtabar, H., & Saidi-Mehrabad, M. (2019). *Cost Engineering and Pricing in Autonomous Manufacturing Systems*. Emerald Publishing Limited.

Fazlollahtabar, H., & Shafieian, S.H. (2014). An optimal path in an AGV-based manufacturing system with intelligent agents. *Journal of Manufacturing Science and Production*, 14(2), 87–102.

Fazlollahtabar, H., Es'haghzadeh, A., & Hajmohammadi, H., et al. (2012). A Monte Carlo simulation to estimate TAGV production time in a stochastic flexible automated manufacturing system: A case study. *International Journal of Industrial and Systems Engineering*, 12(3), 243–258.

Fazlollahtabar, H., Mahdavi-Amiri, N., & Muhammadzadeh, A. (2015). A genetic optimization algorithm for nonlinear stochastic programs in an automated manufacturing system. *Journal of Intelligent & Fuzzy Systems*, 28(3), 1461–1475.

Fazlollahtabar, H., Rezaie, B., & Kalantari, H. (2010). Mathematical programming approach to optimize material flow in an AGV-based flexible jobshop manufacturing system with performance analysis. *The International Journal of Advanced Manufacturing Technology*, 51(9–12), 1149–1158.

Fazlollahtabar, H., Saidi-Mehrabad, M., & Balakrishnan, J. (2015a). Mathematical optimization for earliness/tardiness minimization in a multiple automated guided vehicle manufacturing system via integrated heuristic algorithms. *Robotics and Autonomous Systems*, 72, 131–138.

Fazlollahtabar, H., Saidi-Mehrabad, M., & Balakrishnan, J. (2015b). Integrated Markov-neural reliability computation method: A case for multiple automated guided vehicle system. *Reliability Engineering & System Safety,* 135, 34–44.

Fazlollahtabar, H., Saidi-Mehrabad, M., & Masehian, E. (2015). Mathematical model for deadlock resolution in multiple AGV scheduling and routing network: A case study. *Industrial Robot: An International Journal,* 42(3), 252–263.

Fazlollahtabar, H., Saidi-Mehrabad, M., & Masehian, E. (2021). Robotic industrial automation simulation-optimization for resolving conflict and deadlock. *Assembly Automation,* 41(4), 477–485.

Kouvelis, P., & Yu, G. (1997). *Robust Discrete Optimization and its Applications.* Kluwer Academic Publishers, Boston.

Montemanni, R., Gambardella, L.M., & Donati, A.V. (2004). A branch and bound algorithm for the robust shortest path problem with interval data. *Operation Research Letters,* 32, 225–232.

Nishi, T., Hiranaka, Y., & Grossmann, I.E. (2011). A bilevel decomposition algorithm for simultaneous production scheduling and conflict-free routing for automated guided vehicles. *Computers & Operations Research,* 38, 876–888.

Shiazi, B., Fazlollahtabar, H., & Mahdavi, I. (2010). A six sigma based multi-objective optimization for machine grouping control in flexible cellular manufacturing systems with guide path flexibility. *Advances in Engineering Software,* 41(6), 865–873.

Shojaeifar, A., Fazlollahtabar, H., & Mahdavi, I. (2016). Decomposition versus minimal path and cuts methods for reliability evaluation of an advanced robotic production system. *Journal of Automation Mobile Robotics and Intelligent Systems,* 10(3), 52–57.

Siddiqui, A.W., & Verma, M. (2015). A bi-objective approach to routing and scheduling maritime transportation of crude oil. *Transportation Research Part D,* 37, 65–78.

Tavana, M., Fazlollahtabar, H., & Hassanzade, R. (2014). A Bi-Objective stochastic programming model for optimizing automated material handling systems with reliability considerations. *International Journal of Production Research* 52(19), 5597–5610.

Waller, S.T., & Ziliaskopoulos, A.K. (2002). On the online shortest path problem with limited arc cost dependencies. *Networks,* 40(4), 216–227.

3 Sustainability and Automation

3.1 INTRODUCTION

The Circular Economy (CE) conceptualizes an envisioned global economy which is "restorative and regenerative by design" (Ellen MacArthur Foundation, 2013) and which simultaneously considers environmental impact, resource scarcity, and economic benefits (Lieder and Rashid, 2016). In practice, CE-related issues – such as the reduction of waste, emissions, and supply risk – are spurring innovations in business models, product designs, materials, and supply chain configurations. Circular business approaches focus on "maintaining the highest level of economic value of products, components and materials for as long as possible, while at the same time ensuring that the environmental impact over time is as low as possible" (Balkenende et al., 2017). In a CE, businesses need to find ways to make profit from "the flow of materials and products over time" in a system where products and materials are continually reused (Bocken et al., 2016). In particular, service-oriented business models, such as Product Service Systems (PSS), are often mentioned for their potential to reduce environmental impacts (Qu et al., 2016), and support the CE (Tukker, 2015). PSS have been defined as innovation strategies that "shift the business focus from designing and selling physical products only, to designing and selling a system of products and services that are jointly capable of fulfilling specific client demands" (Manzini, E.; Vezzoli, 2003). On the product level, circular design strategies prescribe a lifecycle perspective, targeting product features such as durability, upgradability, reparability, and recyclability.

The Internet of Things (IoT) has been described as a new paradigm in which everyday objects can sense and communicate (Atzori et al., 2010) leading to completely new possibilities for information exchange (Whitmore et al., 2015). The IoT has been defined as a "system of uniquely identifiable and connected constituents capable of virtual representation and virtual accessibility leading to an Internet-like structure for remote locating, sensing, and/or operating the constituents with real-time data/information flows between them, thus resulting in the system as a whole being able to be augmented to achieve a greater variety of outcomes in a dynamic and agile manner" (Ng and Wakenshaw, 2017). The IoT brings about the possibility to collect large amounts of data from products in use. The business opportunities

DOI: 10.1201/9781003400585-3

of the IoT are thus linked to developments in other technologies such as real-time computing, machine learning, and big data analytics (Stankovic, 2014). The implementation of IoT in a company can support real-time data processing and optimized resource use which could lead to the development of more competitive products and more profitable business models (Li et al., 2015). As the IoT takes form and expands, with a growing number of smart and connected products, physical objects are increasingly able to understand and react to their environment (Kortuem et al., 2010). This allows for improved visibility of assets in the field, with implications for a CE. For example, manufacturers can gain knowledge about the current and predicted condition of products, and thereby build services based on actual performance and use (Baines and Lightfoot, 2013). Moreover, connecting products to the IoT can support monitoring of products and parts throughout their lifecycles, and provide decision support for companies implementing circular business models.

Previous research has categorized the opportunities of the IoT for a CE into "smart" maintenance, reuse, remanufacturing, and recycling (Alcayaga et al., 2019). However, publications covering empirical work, and in particular case study research, are still limited (Nobre and Tavares, 2017). Specifically, extant literature does not give an answer to how current practice compares to the envisioned opportunities of the IoT for the CE.

In this chapter, we therefore aim to better understand how the IoT is currently implemented for the CE in practice, and how that compares to the literature about how the IoT might support the CE. To this end, we aim to answer the following research question:

How have companies to date implemented the IoT for circular strategies and how are these implementations distributed between the opportunities anticipated in literature?

To answer this question, we also pose the following sub-question:

How can IoT-enabled circular strategies be categorized in a framework which enables mapping a variety of company cases?

3.2   RELATED WORKS

Existing literature about the IoT within the context of the CE is sparse but growing (Nobre and Tavares, 2017). While there is a large pool of literature available in the fields of the IoT and the CE independently, there is still limited research published in the nexus between the two (Tseng et al., 2018), and there is a need for more research to systemically map CE approaches to emerging digital technologies (Okorie et al., 2018). Below, we summarize previous work that discusses how the IoT could be used throughout the lifecycle of a product to reduce environmental impact, prolong product lifetimes, and close material loops.

In a review paper on digital technologies in the circular economy, Pagoropoulos et al. (2017) noted that the IoT can enable monitoring of the health and actions of connected products. Salminen et al. (2017) discussed that increasing intelligence and automation can create new business opportunities and help optimize existing operations that are favorable in a circular economy, while Spring and Araujo (2017) argued that "smart products" allow for "connected, rich biographies of products" which can support activities such as maintenance and reverse logistics, especially when products "circulate beyond the direct governance of one coordinating firm". Jensen and Remmen (2017) similarly found that digitalization could potentially support product lifecycle management and the integration of information about, for example, material composition of products, which could stimulate high-quality recycling and reuse. Roy et al. (2016) emphasized the role of "life cycle "big data" analytics" for continuous maintenance of products". De Sousa Jabbour et al. (2018) discussed the relationship between the Circular Economy, Industry 4.0, and sustainable operations management. They suggested, for example, that "product passports" can improve recovery, and that tracking of products can enable reverse logistics. Gligoric et al. (2019) highlighted the lack of a unified ontology for data exchange to support circular strategies, and presented a technology for printing sensors that could serve as product passports. These passports would carry data about, for example, material composition, recyclability, and potential for reuse.

Only a small number of peer-reviewed case studies have been published that describe the use of the IoT for improved circularity of products and services in practice. Lightfoot et al. (2011) interviewed representatives from four companies about how ICT can support the implementation of advanced services. They provide examples of how data from connected trucks can support optimal fuel efficiency, and how data about the location and condition of trains can enable effective and efficient maintenance services. In the context of building equipment, Fargnoli et al. (2019) highlighted how building information management systems can support more effective management of maintenance activities and enhance information exchange between stakeholders in the equipment lifecycle. Grubic and Jennions (2018) performed a multiple case study in which they extracted factors that "characterize the application of remote monitoring technologies in the context of servitised strategies". They found that remote monitoring technology supports "a broad spectrum of product and service combinations, from warranty to availability contracts", and that a complex relationship exists between the technology and the servitized strategies. Ardolino et al. (2016 and 2018) built on a literature review and a multiple case study to identify key digital capabilities for service transformation. In the context of performance-based contracts they found that, the digital capabilities of "usage monitoring" and "prediction" could support the delivery of equipment uptime, and that the capability for "adaptive control" could support services that promise specific levels of efficiency in product use. Lindström et al. (2017) conducted a case study about recycling management optimization supported by IoT technology. They studied how a company transformed to a PSS provider, how they used the IoT, and how the new set-up affected their customers. Kiritsis (2011) studied the implementation of the IoT for "closed-loop product lifecycle management" in ten demonstrator projects, for example, describing how the IoT was used to identify and assess the condition of vehicle components available for reuse. Moreno et al. (2017) described a case in which a 3D-printed shoe with integrated sensors could alert the user when repair was needed.

Two frameworks have previously been published that give an overview of the enabling effects of digitalization on circular strategies in business. Alacayaga et al. (2019) reviewed literature on the intersections between the IoT, PSS and CE and proposed a concept of "smart-circular PSS" including "smart" use, maintenance, reuse, remanufacturing, and recycling. The authors give some indication about how different strategies are currently implemented in practice, stating that "smart use" and "smart maintenance" have high usage, while "smart reuse" has medium usage, "smart remanufacturing" low usage, and "smart recycling" very low usage. However, this assessment is not clearly based on a review of cases from practice.

Bressanelli et al. (2018) conducted a literature review and a single case study to identify "usage-focused business model functionalities" that could be supported by the IoT and big data analytics, and which had impact on circular strategies. The eight functionalities extracted were the following: improving product design, attracting target customers, monitoring and tracking products activity, providing technical support, providing preventive and predictive maintenance, optimizing the product usage, upgrading the product, and enhancing renovation and end-of-life activities. The framework was designed to analyze a single company case according to IoT-enabled circular strategies applied. The frameworks do not explicitly distinguish between *types* of IoT solutions, which limits their potential for using the frameworks to identify additional opportunities in that dimension. Moreover, the frameworks were not designed to facilitate mapping and comparison between different cases.

To fulfil the aim of this chapter, to better understand how the IoT is currently implemented for CE in practice, and how that compares to the literature about how the IoT might support the CE, we develop a framework which complements previous work by presenting a structured way of categorizing IoT-enabled circular strategies, and which facilitates mapping of diverse cases according to the IoT capabilities used, as well as the circular strategies enabled. We subsequently use this framework to map a larger set of cases from practice in order to provide insights into how current implementation in practice is distributed between the different IoT-enabled circular strategies.

3.3 SUSTAINABLE PRODUCTION WITH ROBOTICS

For years, robots have been considered drivers of productivity and efficiency. With the latest technologies and robot types, the degree of automation in the manufacturing industry is continuously increasing. Industrial robot support allows complex, tiresome, and dangerous work to be automated and performed quickly. The tasks range from vacuum cleaning robots to bomb defusers. In industrial manufacture, robots can be used for point automation or integrated system solutions. Although companies are using robotics and automation with increasing frequency, they have so far rarely made the leap to fully integrating complex robot systems. Machinery that appears to come from a science fiction film increases productivity and reduces costs, but has other major advantages that are often overlooked: sustainability and energy efficiency.

Robots in manufacture represent a great opportunity for improving energy and resource consumption and for circular recycling. From a social point of view, robots can have positive effects on individual production processes thanks to human-machine collaboration. In robot design, many modern manufacturers use lightweight components, ensuring that as many of them as possible can be recycled. Another factor that makes robot technology sustainable is its durability. Many robot manufacturers make it possible to safely use robust robots that require no maintenance or regular lubrication. This is the only way to improve technology and production processes while keeping investment low.

Sustainability has become a major component of competition among manufacturers worldwide. Approximately one-third of the energy consumed in the United States is used for manufacturing. Sustainable production is the top concern for manufacturers and consumers because generating greenhouse gases and harmful emissions and disposing of waste generated during the manufacturing processes could pollute the environment. Therefore, production system design, analysis, and control play key roles in providing cleaner energy and reducing emissions, waste, and pollution, as well as minimizing the impact of climate change and conserving natural resources and energy during the product's entire life cycle. Therefore, Sustainable Production Automation (SPA), which involves designing, structuring, and engineering operations and products, is a significant part of manufacturing.

3.3.1 Reduction of Energy Costs

A major incentive for automated manufacture is energy cost reduction. The modern technologies of the latest robot generations allow completely integrated manufacturing processes developed specifically for automation with regard to the energy they consume. Energy efficiency results from the optimisation of individual manufacturing processes, which taken together accelerates production. Robots work without interruption and perform several tasks in a single step. You do not need any light, heating or direct supervision, which results in significant savings in energy costs.

3.3.2 Minimizing Production Waste and Material Consumption

Cutting-edge technologies rely on extremely high-precision movements. This minimizes errors and optimizes the planning of necessary production material. This in turn minimizes production rejects and excess material. For instance, in automotive production, the amount of adhesive or paint required for an individual component can be calculated and programmed exactly.

3.3.3 Replacing Parts with Robots instead of Large Systems

Robots can be programmed in a wide variety of ways, making them much more flexible than static machines. Sticking with the example of automotive production, replacement parts continue to be needed years after a series has stopped being produced. The manufacturer can use a modular robot system to simply retrofit the production process. Large systems that require a great deal of energy and space and a large number of employees can be eliminated.

3.3.4 SUPPORT FOR THE RECYCLING PROCESS

Robots can help close the circuit that consists of production, sorting and recycling of used goods. Integrated vision systems allow them to detect individual components and disassemble them into recyclable parts.

3.3.5 HUMANS AND MACHINES

The question now is how well robots can be integrated from a social point of view. According to a study conducted by the Massachusetts Institute of Technology (MIT) in co-operation with BMW, the motor vehicle manufacturer, mixed groups of robots and humans are about 85% more productive than teams made up only of robots or only of humans.

As has been mentioned, robots can also perform a number of tiresome, repetitive, or even dangerous tasks, relieving stress on humans. This creates capacity for more productive tasks for employees.

3.4 METHODOLOGY

To achieve the aim of this chapter, we first developed a framework to categorize IoT-enabled circular strategies, and then used the framework to map a set of cases from practice. The research was carried out in four main steps: (1) rapid literature review and framework development; (2) identification of cases; (3) selection of relevant cases; and (4) analysis and mapping of cases to the framework.

3.4.1 RAPID LITERATURE REVIEW AND FRAMEWORK DEVELOPMENT

In order to categorize the company cases, a framework was developed distinguishing between different categories of IoT capabilities on the one hand, and different categories of circular strategies on the other hand. This was to identify the key capabilities underpinning the IoT and the key strategies applied in the CE. The methodology was considered appropriate given the focused search target, the small number of academic sources specific to IoT-enabled CE, and the broad scope of the fields of IoT and CE as such. The goal of the review was to identify an established view among scholars about IoT capabilities and circular strategies, respectively. In the field of IoT, papers were collected from academic journals based on the following criteria: the papers should (1) provide an overview of the technology and its enabling capabilities; (2) be general in the sense that the results could be applied across industries; and (3) be well cited (> 15 citations) to demonstrate that the findings were recognized by other researchers.

In the field of the CE, academic literature was consulted to identify existing frameworks describing product and business model design strategies. Similar selection criteria were applied. The papers should (1) explicitly focus on deriving design strategies for CE, and (2) present results that could be applied generally (not case or industry specific), and (3) have more than 15 citations.

The IoT capabilities and circular strategies described in the literature were interpreted and grouped together into categories, using frameworks from "grey" literature as starting points. In the IoT dimension, we started from the four categories: monitoring, control, optimization and autonomy. We then used academic literature sources to extract additional IoT capabilities.

3.4.2 IDENTIFICATION OF CASES FROM PRACTICE

Since there are few case studies published in academic literature about how the IoT can be used for the CE, "grey" literature was used to extract cases from practice. Reports were scanned based on the following criteria: (1) it should discuss the application of IoT technology, while also considering environmental aspects; (2) it should review a large set of cases (> 20), to allow for comparable case descriptions; and (3) it should not be directly authored by the ICT industry, to avoid bias. Based on these criteria, two reports were identified as suitable sources for a comprehensive overview of relevant company cases. One of the reports originates from the field of CE. Both organizations are well known knowledge sources in their respective fields, and were judged to be reliable sources for the case review. Past works present a combined total of 92 cases, describing a wide range of IoT and CE implementations in business. Considering the high number of cases reviewed in the two reports and the broad scope taken by both organizations in assembling the collections of cases, we treated the combined set of cases, although certainly not complete, as a representative sample of current practice in using the IoT for the CE in business.

3.4.3 SELECTION OF RELEVANT CASES

From the 92 cases presented, we selected 40 cases to analyze in this study. The selection process of going from 92 cases to 40 cases. The following selection criteria were applied: (1) the case should be described in enough detail to allow for further analysis; (2) the case should depict the use of a circular strategy; (3) the case should be centered around the use of an IoT-enabled product; and (4) the case should describe an actual implementation to date. 22 cases were excluded because we lacked information to do the subsequent analysis and mapping. For example, if a large manufacturer was mentioned without any description regarding the product or service considered, the case could not be further analyzed. The remaining 70 cases were briefly analyzed in terms of relevance to the IoT and the CE. 12 cases were excluded because they did not describe any impact in terms of the CE or environmental sustainability. Examples of cases that were excluded based on this criterion are FitBit (captures data about daily activities to support a healthy lifestyle) and iRobot Roomba (autonomous vacuum cleaner that uses sensors to navigate). Nine cases were excluded because they did not show a strong enough IoT component, namely, they did not include any interactions with a physical smart product. Examples here are AirBnB (an online platform to match supply and demand for short stay rental apartments) and PayByPhone (an app that makes it easier to pay for parking). Finally, since the focus of the chapter was to understand current implementation, we excluded prototypes and start-ups. Nine cases were excluded based on this criterion, for example Aganza (a platform for offering products using a "pay-as-you-go" model) and Burba (a smart bin prototype).

3.4.4 ANALYSIS AND MAPPING OF CASES

For each of the selected cases, we read the description of the case in the publication where it was presented and highlighted information about the use of the IoT, and about the implementation of circular strategies. When needed, additional data was retrieved from other sources, such as the company's website. Based on the data retrieved, we could map the case to the framework according to the definitions of IoT capabilities and circular strategies. It should be noted that each case can display the use of several IoT capabilities and circular strategies.

3.5 FRAMEWORK

This section describes how the framework was developed based on literature. We explain, separately, how we derive five categories in the IoT dimension and six categories in the CE dimension.

3.5.1 IoT CAPABILITIES

While the exact terminology to describe IoT-related business opportunities varies between authors, we here synthesize core capabilities of the technology as found in literature.

There are four levels of the capabilities of "smart, connected products": monitoring, control, optimization, and autonomy. Monitoring relates to the ability of a product to provide information about its own use, and to sense its environment. For example, monitoring can be applied for diagnostics and prognostics of products-in-use, in order to enhance maintenance services in PSS (Grubic, 2014). On the level of the business model, monitoring can support business models that are based on product access and/or performance, as pricing can be based on actual product usage. Control refers to the ability to change product and system parameters. When a product is connected to the IoT, this control can happen remotely (Gubbi et al., 2013). Moreover, if the product is equipped with the capabilities of monitoring and processing, it can control itself based on insights from monitoring conditional parameters. Optimization can be described as the application of algorithms and analytics to in-use or historical data, to optimize output parameters, utilization, or efficiency. Autonomy brings about an additional layer of self-coordination to the functionality of smart products, giving enhanced ability to control system complexity, still using the capabilities of monitoring, control, and optimization (Atzori et al., 2016).

3.5.2 CIRCULAR STRATEGIES

The CE is still a relatively new term in the academic literature, and a large number of different definitions have been published. Previous research has shown that academics do not fully agree about which aspects should be included in the CE concept, and which not. For the purpose of this chapter, however, we choose to focus on design and business model strategies that can support environmental benefits in the CE. As we focus on IoT enabled products, we limit our search to strategies in the so-called technical metabolism, namely, products that have the potential to be recovered and reused, partly or as an entity, through several life cycles. The Ellen MacArthur foundation (2013) defines five main circular strategies: share, maintain/prolong, reuse/redistribute, refurbish/remanufacture, and recycle. Sharing of products has the potential to increase their utilization, which could reduce the total number of products since one product could satisfy many peoples' needs for a certain function. Maintaining products and prolonging their lifetimes are mentioned as a circular strategy by, e.g., (Den Hollander et al., 2017). A core aspect of the CE is the looping of resources from a post-use stage back to production. Remanufacturing and recycling have been researched extensively in the design for CE literature (e.g., (Bakker et al., 2014; Moreno et al., 2016)). Strategies for the reuse of products are also often described in the CE literature, sometimes framed as part of product-lifetime extension strategies and sometimes as part of looping strategies. From a broader perspective, a core aspect of circular strategies is to support systemic change towards a more sustainable economy. In the design literature, previous research has produced specific design tools that aim to facilitate systems change. Such tools support designers in applying "whole systems design", considering the complex system around the product or service in order to ensure environmental and societal benefits (Moreno et al., 2016).

3.5.3   Final Framework

Building on the IoT capabilities and the circular strategies derived from literature, we propose a framework composed as a matrix in which the categories of IoT capabilities and circular strategies form the columns and rows respectively. Each combination of IoT capability and circular strategy forms a IoT-CE cross-section as shown in Figure 3.1. Tracking relates to the ability to uniquely identify and localize assets. Monitoring describes the use of sensors and metering devices to give information about a product's use, condition, and environment. Control makes it possible to steer product operations through some type of digital interface. Optimization relates to goal-based improvement using advanced algorithms as well as monitoring and/or control capabilities. Optimization can be applied to multiple levels. It can refer to adapting product-internal operations, system operations, or service operations such as maintenance. Finally, design evolution is the ability to learn from product-in-use data in order to improve the design of a product or a service. We note that this capability is different from the others in the way that it does not necessarily change the functionality of the product itself. Instead, it allows for learning about in-the-field product parameters, such as performance and use patterns. These insights can then be used in the design of a next generation of the product or PSS, rather than during use.

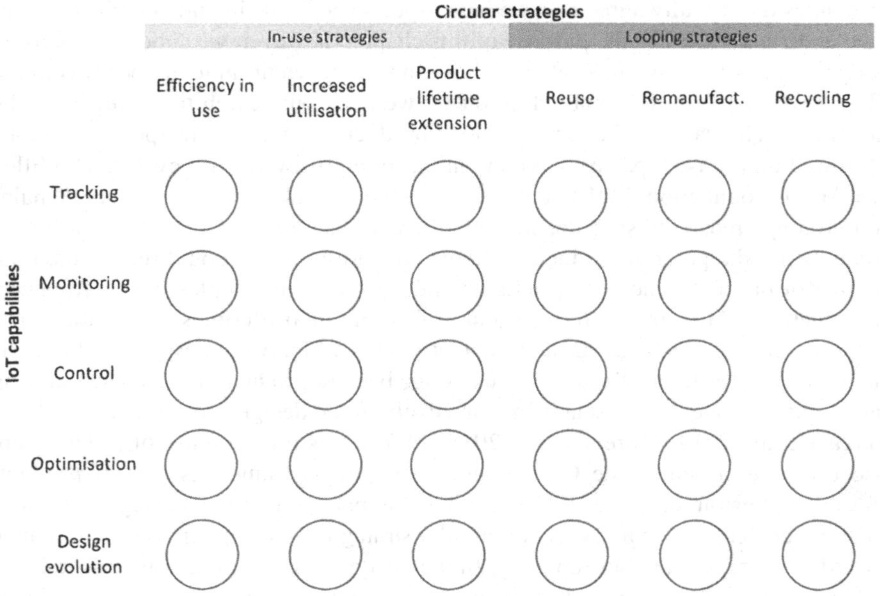

		Circular strategies			
	In-use strategies			Looping strategies	
	Efficiency in use	Increased utilisation	Product lifetime extension	Reuse	Remanufact.	Recycling

IoT capabilities

Tracking ◯ ◯ ◯ ◯ ◯ ◯

Monitoring ◯ ◯ ◯ ◯ ◯ ◯

Control ◯ ◯ ◯ ◯ ◯ ◯

Optimisation ◯ ◯ ◯ ◯ ◯ ◯

Design evolution ◯ ◯ ◯ ◯ ◯ ◯

FIGURE 3.1   Framework categorizing IoT capabilities and circular strategies.

3.6   MAPPING THE CASES TO THE FRAMEWORK

The 40 analyzed cases cover a wide range of products and industries, targeting different customer groups. High-investment products such as mining equipment, wind turbines, and jet engines are examples of products represented in the cases. Moreover, assets in the built environment, infrastructure and the power grid are represented, as are smart connected cars used in different types of car sharing services. Several cases describe solutions for energy management in buildings through smart products such as thermostats, fans, HVAC systems, elevators, washing machines, and lighting systems. Printers and ATM machines are also represented. Most of the identified cases show products and services offered in a business-to-business context, but business-to-customer and business-to-government examples are also present in the set. All cases concern products that can be looped through the technical cycle, namely, that have the potential to be recovered and reused through several life cycles.

Figure 3.2 shows the mapping of the 40 company cases to the framework. Each case was analyzed and mapped according to the categories of IoT capabilities and circular strategies, resulting in a "heat map" of the occurrences of IoT-CE cross-sections. Many of the cases describe more than one IoT-CE cross-section, leading to 150 categorized occurrences of IoT-CE cross-sections from the 40 cases. It should further be noted that while the categories are uniquely defined, they are not mutually exclusive. For example, cases displaying optimization often also rely on the use of monitoring or control capabilities. The results show that examples of IoT-enabled "efficiency in use" and "product lifetime extension" dominate in the analyzed set of cases, while IoT-enabled looping and design evolution are relatively unexplored.

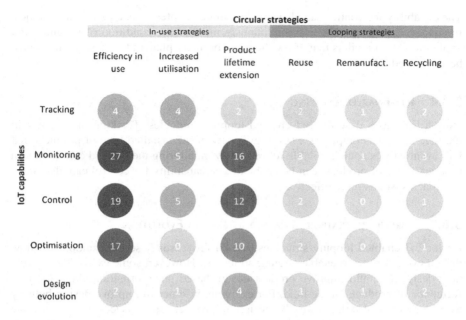

FIGURE 3.2   Heat map of the IoT-CE cross-section occurrences.

3.6.1 IoT-Enabled Efficiency in Use

The cases in this group display the use of IoT for different types of efficiency measures in the use phase. Many examples describe strategies for saving energy, for example through tracking of vehicles, or by monitoring energy use and performance. Moreover, remote control of energy consuming products allows users to save energy, while advanced energy management systems can learn from the user's routines to optimize the system for efficiency and comfort.

3.6.2 IoT-Enabled Increased Utilization

Several cases in this group describe services that allow users to access and use products that they do not own. For example, in a car-sharing service, the IoT allows users to find a car that are close to and which is available for use. The service provider can store information about who has used the car, for how long, and at what time. Another example is increased utilization of space in office buildings, allowing fewer rooms to serve more people.

3.6.3 IoT-Enabled Product Lifetime Extension

Cases in this group describe maintenance and repair activities, as well as upgrades. For example, tracking allows maintenance operatives to identify and locate products that need to be serviced. Moreover, products can monitor their own use and status, and send alerts about when they need maintenance, and which spare parts to order.

The capability for control can also enable remote maintenance, repair, and upgrades. In advanced cases, companies continuously monitor the condition of products and apply prediction models that allow them to optimally plan and execute maintenance before a product fails.

3.6.4   IoT-Enabled Looping

Only a few cases describe IoT-enabled looping strategies. Tracking of products in the field can give companies a better overview of the installed base of products that will eventually become available for reuse. By monitoring the use and condition of a product, reuse of products can be enabled. The capability for control can also make take-back systems more effective.

3.6.5   Circular Strategies Enabled by Design Evolution

As for IoT-enabled looping strategies, only a few cases describe examples of how design evolution could enable circular strategies. However, some examples describe how designers could improve product design based on information about the performance of products in the field. For example, in order to improve design both by avoiding technical failures, and by designing products and services that users value over time, designers could use detailed information about how products are used and discarded.

3.7   DISCUSSION

Based on our results, we further note that the capability for design evolution – the feedback of product-in-use data to design in order to improve products and services – is not commonly used in practice to support circular strategies. To the best of our knowledge, this result has not been reported previously. By pointing out underexplored areas, the findings in this chapter provide a starting point for future research into why IoT-enabled looping and design evolution for circular strategies are not implemented to scale. We suggest future research into the real-world challenges of implementing circular looping strategies, and the opportunities and challenges of using IoT technology to support such strategies and associated business models. Finally, some limitations of the chapter in terms of scope and methodology should be mentioned. In this study, we intentionally focused on reviewing implemented solutions, namely, solutions that are currently offered to customers. Therefore, we excluded prototypes and start-up companies from the analysis. Analyzing such early stage initiatives would enrich the picture with possible future developments in the implementation of IoT-enabled circular strategies. Moreover, actual assessment of impact on environmental sustainability was outside the scope of this chapter. In order for practitioners to be able to effectively use our framework as a tool when innovating IoT-enabled circular strategies, it should be complemented with other tools that provide information on the sustainability impact of IoT-enabled circular solution. From a methodology point of view, our study is limited in that it is based on two reports to source the company cases. Although the 40 selected cases include a diverse range of companies and products, it is possible that some relevant examples were not covered in the set. Thus, while the results presented in this chapter answer the research question posed in Section 3.1 by presenting the distribution of implementations in the 40 cases, future studies could include additional cases in order increase the robustness of the results.

Also, the mapping of the cases to the framework unavoidably involved a degree of interpretation. However, we argue that although some cases could have been interpreted differently, it would not significantly change the main result – namely, the relative distribution in the implementation of the different IoT-enabled circular strategies. Also, as we show how we categorize each case, it is possible for other researchers to analyze the same set of cases and potentially come up with an alternative categorization. Lastly, the rapid literature review approach followed when developing the framework is limited in that it is less comprehensive than a full "systematic literature review" FRANCIS. However, the goal of this review was not to compile a comprehensive list of publications in the fields of IoT and CE, but to identify commonly agreed views among scholars on IoT capabilities and circular strategies, respectively. Among the criteria used for selecting papers to include in the rapid literature review, one was to omit papers with few citations. This criterion was chosen since a high number of citations was considered a suitable indication that the results were accepted and acknowledged by the scientific community. It is possible, however, that this led to the exclusion of relevant articles that have only recently been published, and therefore lack citations at this point in time.

3.8 CONCLUSIONS

This chapter aimed to understand how companies have implemented IoT for circular strategies, and how those implementations compare to anticipated opportunities described in literature. To that end, a two-step approach was followed. First, a framework was develop based on previous literature. The framework gives a categorization of IoT-enabled circular strategies according to the IoT capabilities used, and the circular strategies enabled. The framework complements previously published frameworks, as it adds additional detail and allows for easy mapping of diverse cases. In its current form, the framework could be useful for companies since it provides an overview of IoT-enabled circular strategies. However, the usefulness of the framework for practitioners has not been evaluated in this chapter. Secondly, we collected 40 cases from practice depicting current implementation of IoT for CE, and mapped them to the framework. This way, we could provide concrete examples of cases mapped to different parts of the framework, and derive practice-based insights about the current distribution of implementation of IoT-enabled circular strategies. Our results show that current implementation of IoT-enabled circular strategies mainly supports "efficiency in use" and "product lifetime extension". Only a small number of the reviewed cases displayed IoT-enabled "looping" (reuse, remanufacturing, and recycling). This is a notable result, as "closing the loop" is one of the main goals expressed in the CE literature. Moreover, few cases described "design evolution" for CE, namely, using the feedback of data from products-in-use to improve circular design.

Future research could build on the results presented in this chapter to develop a tool for companies to map their current state of IoT-enabled CE implementation, and to improve their strategy further. To be able to provide such guidance for practitioners, more studies are needed. Specifically, future research should investigate *why* IoT-enabled looping strategies or design evolution for circular strategies have not been implemented to scale. To do so, in-depth case studies with companies would be relevant, as such studies could extract context-specific opportunities as well as implementation barriers perceived by the companies. Moreover, insights from studies focusing on the actual sustainability impact of different IoT-enabled circular strategies would be important as input to recommendations for practitioners about how to reap the benefits of IoT for CE.

BIBLIOGRAPHY

Alcayaga, A., Wiener, M., & Hansen, E.G. (2019). Towards a framework of smart-circular systems: An integrative literature review. *Journal of Cleaner Production*, 221, 622–634.

Ardolino, M., Rapaccini, M., Saccani, N., et al. (2018). The role of digital technologies for the service transformation of industrial companies. *International Journal of Production Research*, 56, 2116–2132.

Ardolino, M., Saccani, N., Gaiardelli, P., et al. (2016). Exploring the key enabling role of digital technologies for PSS offerings. *Procedia CIRP*, 47, 561–566.

Atzori, L., Iera, A., & Morabito, G. (2010). The Internet of Things: A survey. *Computer Networks.*, 54, 2787–2805.

Atzori, L., Iera, A., & Morabito, G. (2016). Understanding the Internet of Things: Definition, potentials, and societal role of a fast evolving paradigm. *Ad Hoc Networks.*, 56, 122–140.

Baines, T., & Lightfoot, H. (2013). Information and Communication Technologies. In *Made to Serve: How Manufacturers Can Compete through Servitization and Product-Service Systems*. Baines, T., Lightfoot, H., Eds.; John Wiley & Sons Incorporated, Chichester, UK, pp. 169–180.

Bakker, C., Wang, F., Huisman, J., et.al. (2014). Products that go round: Exploring product life extension through design. *Journal of Cleaner Production*, 69, 10–16.

Balkenende, R., Bocken, N., & Bakker, C. (2017). Design for the Circular Economy. In *The Routledge Handbook of Sustainable Design*. Egenhoefer, R.B., Ed.; Routledge, London, UK, pp. 1–19.

Bocken, N., De Pauw, I., Bakker, C., et al. (2016). Product design and business model strategies for a circular economy. *Journal of Industrial and Production Engineering*, 33, 308–320.

Bressanelli, G., Adrodegari, F., Perona, M., et al. (2018). Exploring how usage-focused business models enable circular economy through digital technologies. *Sustainability*, 10, 639.

De Sousa Jabbour, A.B.L., Jabbour, C.J.C., Godinho Filho, M., et al. (2018). Industry 4.0 and the circular economy: A proposed research agenda and original roadmap for sustainable operations. *Annals of Operation Research*, 270, 273–286.

Den Hollander, M., Bakker, C., & Hultink, E.J. (2017). Product design in a circular economy: Development of a typology of key concepts and terms. *Journal of Industrial Ecology*, 21, 517–525.

Ellen MacArthur Foundation. (2013). *Towards the Circular Economy: Economic and Business Rationale for an Accelerated Transition*. Ellen MacArthur Foundation, London, UK.

Fargnoli, M., Lleshaj, A., Lombardi, M., et al. (2019). A BIM-based PSS approach for the management of maintenance operations of building equipment. *Buildings*, 9, 139.

Fazlollahtabar, H. (2016). Parallel autonomous guided vehicle assembly line for a semi-continuous manufacturing system. *Assembly Automation*, 36(3), 262–273.

Fazlollahtabar, H. (2018a). Lagrangian relaxation method for optimizing delay of multiple autonomous guided vehicles. *Transportation Letters*, 10(6), 354–360.

Fazlollahtabar, H. (2018b). Scheduling of multiple autonomous guided vehicles for an assembly line using minimum cost network flow. *Journal of Optimization in Industrial Engineering*, 11(1), 185–193.

Fazlollahtabar, H. (2019a). An Effective mathematical programming model for production of automatic robot path planning. *The Open Transportation Journal*, 13(1), 11–16.

Fazlollahtabar, H. (2019b). Triple state reliability measurement for a complex autonomous robot system based on extended triangular distribution. *Measurement*, 139, 122–126.

Fazlollahtabar, H. (2020). Comparative simulation study for configuring turning point in multiple robot path planning: Robust data envelopment analysis. *Robotica*, 38(5), 925–939.

Fazlollahtabar, H. (2021). Robotic manufacturing systems using Internet of Things: New era of facing pandemics. *Automation, Robotics & Communications for Industry*, 4.0, 82.

Fazlollahtabar, H. (2022). Internet of Things-based SCADA system for configuring/reconfiguring an autonomous assembly process. *Robotica*, 40(3), 672–689.

Fazlollahtabar, H., Es'haghzadeh, A., Hajmohammadi, H., et al (2012). A Monte Carlo simulation to estimate TAGV production time in a stochastic flexible automated manufacturing system: A case study. *International Journal of Industrial and Systems Engineering*, 12(3), 243–258.

Fazlollahtabar, H., & Hassanli, S. (2018). Hybrid cost and time path planning for multiple autonomous guided vehicles. *Applied Intelligence*, 48, 482–498.

Fazlollahtabar, H., & Jalali, S.G. (2013). Adapted Markovian model to control reliability assessment in multiple AGV. *Scientia Iranica*, 20(6), 2224–2237.

Fazlollahtabar, H., & Mahdavi-Amiri, N. (2013a). Design of a neuro-fuzzy–regression expert system to estimate cost in a flexible jobshop automated manufacturing system. *The International Journal of Advanced Manufacturing Technology*, 67, 1809–1823.

Fazlollahtabar, H., & Mahdavi-Amiri, N. (2013b). Producer's behavior analysis in an uncertain bicriteria AGV-based flexible jobshop manufacturing system with expert system. *The International Journal of Advanced Manufacturing Technology*, 65, 1605–1618.

Fazlollahtabar, H., & Mahdavi-Amiri, N. (2013c). An optimal path in a bi-criteria AGV-based flexible jobshop manufacturing system having uncertain parameters. *International Journal of Industrial and Systems Engineering*, 13(1), 27–55.

Fazlollahtabar, H., Mahdavi-Amiri, N., & Muhammadzadeh, A. (2015). A genetic optimization algorithm for nonlinear stochastic programs in an automated manufacturing system. *Journal of Intelligent and Fuzzy Systems*, 28(3), 1461–1475.

Fazlollahtabar, H., & Niaki, S.T.A. (2017a). Binary state reliability computation for a complex system based on extended Bernoulli trials: Multiple autonomous robots. *Quality and Reliability Engineering International*, 33(8), 1709–1718.

Fazlollahtabar, H., & Niaki, S.T.A. (2017b). Integration of fault tree analysis, reliability block diagram and hazard decision tree for industrial robot reliability evaluation. *Industrial Robot: An International Journal*, 44(6), 754–764.

Fazlollahtabar, H., & Niaki, S.T.A. (2017c). *Reliability Models of Complex Systems for Robots and Automation*. CRC Press.

Fazlollahtabar, H., & Niaki, S.T.A. (2018a). Cold standby renewal process integrated with environmental factor effects for reliability evaluation of multiple autonomous robot system. *International Journal of Quality & Reliability Management*, 35(10), 2450–2464.

Fazlollahtabar, H., & Niaki, S.T.A. (2018b). Modified branching process for the reliability analysis of complex systems: Multiple-robot systems. *Communications in Statistics-Theory and Methods*, 47(7), 1641–1652.

Fazlollahtabar, H., & Olya, M.H. (2013). A cross-entropy heuristic statistical modeling for determining total stochastic material handling time. *The International Journal of Advanced Manufacturing Technology*, 67, 1631–1641.

Fazlollahtabar, H., & Saidi-Mehrabad, M. (2015a). *Autonomous Guided Vehicles: Methods and Models for Optimal Path Planning*. Springer International Publishing, Germany.

Fazlollahtabar, H., & Saidi-Mehrabad, M. (2015b). Risk assessment for multiple automated guided vehicle manufacturing network. *Robotics and Autonomous Systems*, 74, 175–183.

Fazlollahtabar, H., & Saidi-Mehrabad, M. (2015c). Methodologies to optimize automated guided vehicle scheduling and routing problems: A review study. *Journal of Intelligent and Robotic Systems*, 77, 525–545.

Fazlollahtabar, H., & Saidi-Mehrabad, M. (2015d). *Autonomous Guided Vehicles: Methods and Models for Optimal Path Planning*. Springer International Publishing, Switzerland. ISBN 978-3-319-14746-8

Fazlollahtabar, H., & Saidi-Mehrabad, M. (2019). *Cost Engineering and Pricing in Autonomous Manufacturing Systems*. Emerald Publishing Limited.

Fazlollahtabar, H., Saidi-Mehrabad, M., & Balakrishnan, J. (2015a). Mathematical optimization for earliness/tardiness minimization in a multiple automated guided vehicle manufacturing system via integrated heuristic algorithms. *Robotics and Autonomous Systems*, 72, 131–138.

Fazlollahtabar, H., Saidi-Mehrabad, M., & Balakrishnan, J. (2015b). Integrated Markov-neural reliability computation method: A case for multiple automated guided vehicle system. *Reliability Engineering & System Safety*, 135, 34–44.

Fazlollahtabar, H., Saidi-Mehrabad, M., & Masehian, E. (2015a). Mathematical model for deadlock resolution in multiple AGV scheduling and routing network: A case study. *Industrial Robot: An International Journal*, 42(3), 252–263.

Fazlollahtabar, H., Saidi-Mehrabad, M., & Masehian, E. (2021b). Robotic industrial automation simulation-optimization for resolving conflict and deadlock. *Assembly Automation*, 41(4), 477–485.

Fazlollahtabar, H., & Shafieian, S.H. (2014). An optimal path in an AGV-based manufacturing system with intelligent agents. *Journal of Manufacturing Science and Production*, 14(2), 87–102.

Gligoric, N., Krco, S., Hakola, L., et al. (2019). SmartTags: IoT product passport for circular economy based on printed sensors and unique item-level Identifiers. *Sensors*, 19, 586.

Grubic, T. (2014). Servitization and remote monitoring technology: A literature review and research agenda. *Journal of Manufacturing Technology Management*, 25, 100–124.

Grubic, T., & Jennions, I. (2018). Remote monitoring technology and servitised strategies–factors characterising the organisational application. *International Journal of Production Research*, 56, 2133–2149.

Gubbi, J., Buyya, R., Marusic, S., et al. (2013). Internet of Things (IoT): A vision, architectural elements, and future directions. *Future Generation Computer Systems*, 29, 1645–1660.

Jensen, J.P., & Remmen, A. (2017). Enabling circular economy through product stewardship. *Procedia Manufacturing*, 8, 377–384.

Kiritsis, D. (2011). Closed-loop PLM for intelligent products in the era of the Internet of things. *Computer Aided Design*, 43, 479–501.

Kortuem, G., Kawsar, F., Fitton, D., et al. (2010). Smart objects as building blocks for the internet of things. *IEEE Internet Computing*, 14, 44–51.

Li, S., Da Xu, L., & Zhao, S. (2015). The Internet of Things: A survey. *Information System Frontiers*, 17, 243–259.

Lieder, M., & Rashid, A. (2016). Towards circular economy implementation: A comprehensive review in context of manufacturing industry. *Journal of Cleaner Production*, 115, 36–51.

Lightfoot, H.W., Baines, T., & Smart, P. (2011). Examining the information and communication technologies enabling servitized manufacture. *Proceedings of the Institution of Mechanical Engineering. Part B: Journal of Engineering Manufacture*, 225, 1964–1968.

Lindström, J., Hermanson, A., Hellis, M., et al. (2017). Optimizing recycling management using industrial internet supporting circular economy: A case study of an emerging IPS 2. *Procedia CIRP*, 64, 55–60.

Manzini, E., & Vezzoli, C. (2003). A strategic design approach to develop sustainable product service systems: Examples taken from the 'environmentally friendly innovation' *Italian prize. Journal of Cleaner Production*, 11, 851–857.

Moreno, M., Turner, C., Tiwari, A., et al. (2017). Re-distributed manufacturing to achieve a circular economy: A case study utilizing IDEF0 Modeling. *Procedia CIRP*, 63, 686–691.

Ng, I.C.L., & Wakenshaw, S.Y.L. (2017). The Internet-of-Things: Review and research directions. *International Journal of Research in Marketing*, 34, 3–21.

Nobre, G.C., & Tavares, E. (2017). Scientific literature analysis on big data and internet of things applications on circular economy: A bibliometric study. *Scientometrics*, 111, 463–492.

Okorie, O., Salonitis, K., Charnley, F., et al (2018). Digitisation and the circular economy: A review of current research and future trends. *Energies*, 11, 3009.

Pagoropoulos, A., Pigosso, D.C.A., & McAloone, T.C. (2017). The emergent role of digital technologies in the circular economy: A review. *Procedia CIRP*, 64, 19–24.

Qu, M., Yu, S., Chen, D., et al. (2016). State-of-the-art of design, evaluation, and operation methodologies in product service systems. *Computers in Industry*, 77, 1–14.

Roy, R., Stark, R., Tracht, K., et al. (2016). Continuous maintenance and the future–Foundations and technological challenges. *Cirp Annals. Manufacturing Technology*, 65, 667–688.

Salminen, V., Ruohomaa, H., & Kantola, J. (2017). Digitalization and Big Data Supporting Responsible Business Co-Evolution. In *Advances in Human Factors, Business Management, Training and Education*. Kantola, J.I., Barath, T., Nazir, S., Andre, T., Eds.; Springer International Publishing, Switzerland, pp. 1055–1067.

Shojaeifar, A., Fazlollahtabar, H., & Mahdavi, I. (2016). Decomposition versus minimal path and cuts methods for reliability evaluation of an advanced robotic production system. *Journal of Automation Mobile Robotics and Intelligent Systems*, 10(3), 52–57.

Spring, M., & Araujo, L. (2017). Product biographies in servitization and the circular economy. *Industrial Marketing Management*, 60, 126–137.

Stankovic, J.A. (2014). Research directions for the Internet of Things. *IEEE Internet Things Journal*, 1, 3–9.

Tseng, M.L., Tan, R.R., Chiu, A.S.F., et al. (2018). Circular economy meets industry 4.0: Can big data drive industrial symbiosis? Resources, *Conservation and Recycling*, 131, 146–147.

Tukker, A. (2015). Product services for a resource-efficient and circular economy-a review. *Journal of Cleaner Production*, 97, 76–91.

Whitmore, A., Agarwal, A., & Da Xu, L. (2015). The Internet of Things—A survey of topics and trends. *Information System Frontiers*, 17, 261–274.

4 Automation in Industry 4.0

4.1 INTRODUCTION

Robot application in manufacturing systems increased extensively due to their versatility and flexibility. Industrial robots like the Autonomous Guided Vehicle (AGV) are used to transfer materials and products among workstations, production cells, or shops. In an assembly system, the material flow is for either partly assembled products or tools and parts from any two production or processing centers. The robots are controlled through rules issued from a control unit considering the production plan, the master schedule, demand forecasting, the capacity plan, and the path plan and the like. There are several optimization decisions being taken with respect to main objectives, namely the optimal path plan, the optimal production schedule, and so forth. To handle the optimization process it is necessary to plan so that collisions of robots are avoided (Thramboulidis and Christoulakis, 2016).

The scheduling of assembly and the routing of the parts through the assembly shop is a significant problem. Due to the high communication and coordination effort between parts and resources, the AGVs would migrate into the control system. If the final products leave the assembly system, the AGV containers could be used to control the physical AGVs to autonomously control the distribution process of the products (Mourtzis et al., 2016).

AGV (containers) are a specific form of mobile code and the software agents' paradigm. They are active in the way in which they choose to migrate between computers at any time during their execution. This makes them a powerful tool for implementing autonomous parts and sub-assemblies in an assembly system. In general, a mobile agent is able to transport its state from one environment to another, with its data intact, and still being able to perform appropriately in the new environment. AGV control software decides when and where to move next. An AGV container accomplishes this move through data duplication. When the control software decides to move the AGV, it saves its own state and transports this saved state to the next host and resumes execution from the saved state. To sum up, the enabling technologies to realize autonomous products are:

- Identification (e.g. RFID)
- Localization (e.g. RFID reader, Wi-Fi, GPS)

DOI: 10.1201/9781003400585-4

- Communication (e.g. Wi-Fi, UMTS)
- Decentralized data processing (e.g. software agents)
- Sensor networks (e.g. visual sensors).

The combination of autonomous resources on one hand and autonomous parts, sub-assemblies, and products on the other hand will lead to autonomous processes where parts and subassemblies allocate resources and coordinate their assembly by themselves (see Figure 4.1).

Such autonomous processes would lead to highly flexible and self-adaptable assembly systems that could make a variety of customized products and deal with fluctuating demand with only little or even no human interventions (Rymaszewska et al., 2017).

Scheduling of assembly systems is characterized by high complexity (number of orders, variety of products, and variety of resources). The general task of assembly scheduling is the assignment of operations to workstations, allocation of resources, and the building of a schedule. The assembly-scheduling problem is similar to the known job shop scheduling. In autonomous assembly systems, the schedule must be reactive to deal with uncertainty and disruptions. Disruptions should be treated at the system level where they appear. In an assembly control system with reactive scheduling capabilities, the different components cannot be independently programmed since an assembly system is a distributed system and the different workstation programs will run in parallel, exchanging information for synchronization and coordination purposes.

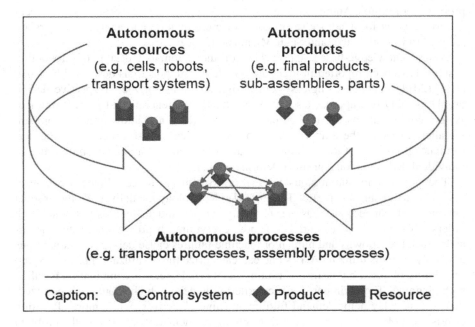

FIGURE 4.1   Intelligent autonomous processes.

Ravi Raju and Krishnaiah Chetty (1993) reviewed design and control issues of flexible manufacturing systems equipped with AGVs. Fazlollahtabar et al. (2010) proposed a flexible job shop automated manufacturing system to optimize the material flow. Xidias and Azariadis (2011) designed mission procedure for multiple AGVs as a general problem. Fazlollahtabar et al. (2012) concerned with applying Tandem Automated Guided Vehicle (TAGV) configurations as material handling devices and optimizing the production time considering the effective time parameters in a Flexible Automated Manufacturing System (FAMS) using the Monte Carlo simulation. Also, Herrero-Pérez and Martínez-Barberá (2010) modeled a distribution transportation system of AGVs for a flexible manufacturing system. Fazlollahtabar and Mahdavi-Amiri (2013a) proposed an approach for finding an optimal path in a flexible job shop manufacturing system considering the two criteria of time and cost using a fuzzy rule backpropagation network. Fazlollahtabar and Mahdavi-Amiri (2013b) proposed a cost estimation model and meanwhile employed a dynamic program to determine the optimal path for the manufacturing network. Fazlollahtabar and Olya (2013) were concerned with proposing a heuristic statistical technique to compute total stochastic material handling time in an Automated Guided Vehicle (AGV) equipped job shop manufacturing system. Fazlollahtabar and Mahdavi-Amiri (2013c) proposed an approach for finding an optimal path in a flexible job shop manufacturing system considering aggregated time and cost under uncertainty. Fazlollahtabar et al. (2015a) considered a scheduling problem for multiple Automated Guided Vehicles (AGVs) in a manufacturing system. They proposed a mathematical program to minimize the penalized earliness and tardiness and developed a heuristic solution approach accordingly. Fazlollahtabar et al. (2015a) proposed a complicated routing/scheduling problem for multiple Automated Guided Vehicles (AGVs) considering a new concept of turning point for deadlock resolution. For more reviews and analysis, readers are referred to Fazlollahtabar and Saidi-Mehrabad (2015b).

A few knowledge-based approaches for the automated design of production systems have been reported in literature. While the support tools described by Khan et al. (2011) generated single system design solutions for their respective design problems, only two approaches have been found that generate and present multiple system configurations for a given production problem, and allow users to compare the alternatives. In these approaches, however, the comparison focuses on the performance properties of the systems (Michalos et al., 2012), or aims at configuring an individual dedicated transfer line (Delorme et al., 2012).

The problem in autonomous production systems is to coordinate the control system so that using Internet of Things (IoTs) technology the delivery of the product on time is guaranteed. The reason is to satisfy demand and prevent lost sales and back orders. Of course, it is essential to handle everything in production system accurately since breakdowns and human or machine failures cause delays. In autonomous assembly systems utilizing AGVs one can categorize the optimization problems into scheduling of assembly tasks and routing of AGVs. Here, this combination is called path planning. The main performance criterion is delay which should be optimized, that is, less delay helps demand fulfillment (time objectives) while more allocation efforts (cost objectives) are required. Applying mathematical modeling, the problem is formulated.

In past works for autonomous assembly systems mainly either cost objectives or time objectives were investigated. Thus, the main contribution of this chapter is to propose a mathematical model to consider delay optimization in an autonomous assembly system using AGVs for material handling. The optimized path plan is a major output of the model helping the decision makers and planners to analyze the material flow and using IoT the data flow in the system.

4.2 PROPOSED PROBLEM AND FORMULATIONS

Originally, the conveyor assembly line was built with multiple workstations and workers with differentiable operating abilities. Various product types are processed on the conveyor assembly line under a given dispatching rule, for example, EDD (Earliest Due Date), FCFS (First Come First Serve), SPT (Shortest Processing Time), and the like. The performance of the traditional manufacturing system is also bound up with the bottleneck process. The bottleneck process increases the make span and yields low motivation amongst skilled workers. Furthermore, the frequent shifting between two different product types results in hard-win and time-consuming setup activities, and a waste of production time and capacity. Consequently, the conveyor assembly line is reconfigured to mitigate or overcome the drawbacks of the traditional manufacturing system.

This chapter is concerned with developing a mathematical model for production planning based on AGV path planning (scheduling and routing). An automated production system is considered where material handling is performed by AGVs between production cells and product processing is done by autonomous robots. AGVs should be assigned to cells to transfer semi-produced products among production cells based on a process plan. The assignments are performed to minimize the waiting time of AGVs and job traffic in production cells. A mathematical model is developed for the assignment problem. The main decisions of the proposed model are obtaining the best routing (path plan) for AGVs and assigning AGVs to jobs in the material handling process. Assembly is performed in production cells by autonomous robots. Different assembly scenarios exist due to the process plan and the of jobs to be processed by robotic machines. To model the uncertain evolution of requirements, a probabilistic scenario model is proposed. A set of nodes O is defined, over a set T of periods. For each node, a probability of realization $P(O)$ is assigned at the beginning of the considered period (t_0). Each scenario node is characterized by a set of production requirements to be guaranteed if the realization of that specific scenario occurs, leading to a tree structure modeling the evolution of the requirements over the time horizon $(t_0, t_1, t_2, \ldots, T)$. The root node represents the current production problem to be addressed and is assumed to be perfectly known.

In detail, the set of products Po to be produced is associated to a scenario O. A volume $d_{p(o)}$ of products in Po must be delivered to the customers, under the hypothesis of an average lot size $lp(o)$. For each product p in P, the assembly process requirements are expressed in terms of autonomous devises like robots and AGVs. The devices required to assemble a part type P are contained in the set $Jp(o)$. Unitary processing times required for each autonomous device are provided in the set $Mj,p(o)$. Furthermore, $Sp(o)$ provides the assembly sequence for each part type, typically requiring multiple robots. Additional non-operational data regarding each robot, dealing with the floor space requirements, investment costs, and depreciation years are also considered.

The inputs of the production planning activity are the set of products that are assembled in the system, the number of available resources, the detailed layout of the system as well as the due dates coming from the customers which have significant impact on the applied production lot-sizes and the operational costs. Production planning is done on a discrete time horizon W, the resolution of the plan is a working shift (w). The objective is to calculate the production lots x_{pwr} respecting the available capacities, cycle (t^m_p) and setup (t^s) time constraints. In the model, setups are expressed with the binary variables z_{pwr} and γ_{pwr}. When assembling a certain product type, a definite amount of devices φ_{jp} is required, and a given amount n_j of devices from each type j are available for use at the beginning of the period. The order demands d_p need to be satisfied by delivering certain amount s_{pw} of products to customers. In the production planning, holding inventory of products (in_{pw}) is allowed, however, it has certain costs c^i. Similarly, planned backlogs (b_{pw}) might occur, but they are also penalized with cost c^b per product and shift. Production and path planning is formulated as an integer programming problem:

Indices:
i Number of jobs ($i = 1, 2, ..., I$)
j Number of autonomous device types ($j = 1, 2, ..., J$)
r Number of production cells ($r, e, s, h = 1, 2, ..., R$)
v Number of AGVs ($v = 1, 2, ..., V$)
m Number of paths ($m = 1, 2, ..., M$)

Other parameters:
P_{ir} Processing time of job i in cells r
WT_{ir} Waiting time of job i in cells r
D_{er} Distance time of cell e to cells r
TD_{erm} Distance time of cell e to cells r from path m

Decision variables:
x_{pwr} Production lots
$Y_{vierm} =$ 1, if AGV v carries job i from cell e to cell r in path m, and 0, otherwise.

Mathematical model:

$$\min \sum_p \sum_w \left(c^b b_{pw} + c^i in_{pw} \right) + \sum_{i=1}^{I} \sum_{r=1}^{R} DT_{ir}. \tag{1}$$

s.t.

$$\sum_r \sum_p \varphi_{jp} \gamma_{pwr} \leq n_j, \qquad \forall w, j, \tag{2}$$

$$\sum_p \left(t_p^m x_{pwr} + t^s z_{pwr} \right) \leq t^p, \qquad \forall w, r, \tag{3}$$

$$d_p \leq s_{pw}, \qquad \forall p, w, \tag{4}$$

$$in_{pw} - b_{pw} = in_{pw-1r} - s_{pw} + \sum_r x_{pwr}, \qquad \forall w, p. \tag{5}$$

$$TD_{erm} = \{(D_{er}), (D_{es} + D_{sr}), (D_{es} + D_{sh} + D_{hr}), \ldots\},$$
$$\text{where, } e \neq r \neq s \neq h \quad e, r, s, h \in shop, \tag{6}$$

$$DT_{ir} \geq \text{Min}(|P_{ir} + WT_{ir} - TD_{erm}|) Y_{vierm}, \qquad \forall\ v, i, e, r, m, \tag{7}$$

$$\sum_{v=1}^{V} \left(Yvierm + Yvirem \right) \leq 1, \qquad \forall\ i, r, e, m \ \text{ and } e \neq r, \tag{8}$$

$$Y_{vierm} \in \{0, 1\}, \qquad \forall\ v, i, r, e, m. \tag{9}$$

The objective function (1) of the problem is to minimize the total backlog and inventory costs that incur in the period and the delay time (DT_{ir}) of AGVs to production cells to handle the jobs.

The first constraints include the limited amount of devices (2) and human capacities (3). Inequality (4) states that demands must be fulfilled, and the balance Eq. (5) links the subsequent production shifts. For the calculation of the setups (z_{pwr} and γ_{pwr}), the multi-item single-level lot sizing model was applied. Constraint (6) shows that to compute the AGV arrival time to a cell, all paths between any two cells are considered as having different time distances. Constraint (7) indicates that the delay time of an AGV carrying job i to production cell r is greater than or equal to the difference between total processing and waiting times, and distance time of cell e to cell r from path m. Constraint (8) certifies that a path is assigned to one AGV at the same time. In relation (9) the binary decision variable is certified.

4.3   GREEDY-BASED ALGORITHMS

The greedy Algorithm is a method that has been studied for dispatching vehicles in a mega Container Terminal. The Greedy Method has been proven as a simplistic heuristic dispatching strategy, and it has earned its recognition as a suitable method for dispatching a single ship and single Quay Crane (QC) situation. The jobs are arranged with a method, called FIFO (First In First Out). In addition to that, the Greedy Algorithm has shown that it also performs reasonably well for multiple QCs and vehicles. However, other studies have shown that the Greedy Method performance deteriorates significantly when it comes to serving multiple QCs. Furthermore, researchers introduced the "refined Greedy Algorithm". The difference between the two is that the refined Greedy Algorithm has a rule. The results from using the "refined Greedy Algorithm" showed an average deviation of 1.55% deviation from the optimal solution for all the problems they conducted.

Another model for dispatching is the Network Flow Method (NFM), which is an extension of an earlier method by another group of researchers. There have been some studies regarding the NFM with the focus on Minimum Cost Flow (MCF). Unfortunately, the MCF was only benchmarked against the greedy method; it would have been interesting to see a comparison of the MCF and the refined Greedy Method that could look ahead.

The MCF had lower waiting time and fewer late jobs. The overall conclusion is that the MCF is optimal for minimizing the waiting time for the AGVs in a container terminal. Yet another study showed similar results that a MCF algorithm out preformed the greedy Algorithm in both AGV waiting time and AGV traveling time. However, that study also showed that greedy had a better statistical deviation when it came to appointment times.

An inventory-based approach has also been used to solve the dispatching problem of AGVs in a Container Terminal. The problem formulation for inventory-based approach was solved with the well-known algorithm, the Hungarian method. They compared the inventory-based approach against the due time-based method which also use the Hungarian method. Finally, they had a modified inventory-based method that had dual cycle time. At first the study indicated that the due time-based methods seemed to allow for more precise scheduling however, when they took bad time estimates into account which is quite common in practice, the inventory-based approach out preformed the due time methods.

The overall conclusion of that study was that the inventory approach was better suited for AGVs in this context. This conclusion was due to the fact that the inventory-based approach does not use the time estimates to the same extent as the time-based ones and this makes it less fragile to bad estimates. Furthermore, it does not require as frequent updates of times, which leads to simpler terminal control.

Another study that has been conducted focused on the relations between the AGVs, cassettes and QCs. Researchers used an open source simulation tool called DESMO-J, which is based on the JAVA programming language. They used a protocol called Contract Net to coordinate the cranes, AGVs and cassettes for loading and discharging containers. The results from their simulations indicated that it is possible to add more AGVs and cassettes instead of adding additional QC with its own set of AGVs and cassettes and maintain or increase the same handling rate. As the cost of one QC exceeds the cost of several AGVs and cassettes, it is not applicable to add more QCs to improve the utilization.

The Greedy Method is using a simple strategy, which in this case is focused on the time. It could be focused on something else or on several different attributes such as fuel consumptions and idle time, which also increases the complexity of the algorithm. The Greedy Method has not been tested with AGVs as a new AGV system has recently been developed. How the Greedy Method would work in reality in a container terminal with AGVs is best explained through a scenario. This is a purely fictional scenario to illustrate how the Greedy Method works.

The first four jobs are not included in the Greedy Method as in this scenario the four AGVs are already queuing at this one crane scenario. However, as the AGVs are queued the results would be the same because AGV 1 is the closest one to the job. The Greedy Algorithm is used from Jobs 5 – 10 with the following description:

- Job 5: Will be delegated to AGV 1 as it has given the lowest time unit for the pickup.
- Job 6: Will be delegated to AGV 3 even though AGV 2 has given the same time unit for the pickup, this is because AGV 3s previous job (Job 5) made it come ahead of AGV 2.
- Job 7: As AGV 2 has given the lowest time unit, it will receive the job.
- Job 8: Again two different AGVs have responded with the same time unit for the pickup of job, this time the AGV 4 was ahead of AGV 1 and therefore it got the job.
- Job 9: Is a clear choice and the job goes to AGV 1.
- Job 10: The final job of this short scenario is delegated to AGV 2.

4.3.1 GREEDY LOOK-AHEAD

The previous scenario illustrated the basics of the normal Greedy Method. There would be some differences when using the Greedy Look-Ahead method. Using the Look-Ahead method raises another question namely: how far should the method look ahead?

After further investigations and a deeper understanding of the different algorithms, we interpreted the Network Flow algorithms to be very similar to the Greedy Look-Ahead algorithm. The research done regarding Network Flow had over 3000 jobs and 10 000 000 arcs which is very time consuming and therefore they limited their choice to the best solution within two minutes. The other study used the Greedy Algorithm, which was able to look ahead, but which did not give an account of how far their method was able to look ahead. Looking ahead through the whole loading/unloading list and dispatching the jobs according to those results would make the dispatching somewhat static and more fragile to discrepancies in the estimated times for the jobs which would also make it harder to make accurate decisions. Therefore, we will find the optimal solution for a smaller number of jobs and calculate every single combination in order to elicit the optimal combination, and of course, the length of the look ahead can be altered. The reason for this approach is to reduce the overall impact on the results if any interruptions or delays should occur. Instead, the impact will mainly affect the set of jobs in that Look-Ahead sequence and the next sequence will get analyzed after the first sequence is completed.

4.3.2 INVENTORY BASED

The Greedy and the Refined Greedy heuristic have been discussed for the purposes of dispatching. Another approach is presented for the dispatching problem, namely inventory based. As mentioned before, the inventory-based solution has shown some promising results especially regarding the AGV and QC waiting time. Mainly, due to the fact that the heuristic of this approach is not built upon due times, instead it is built upon a rough analogy to inventory management. The basic idea of the formulation of the inventory-based method is to avoid estimates of completion times, driving times, due times and tardiness. These estimates are often highly unreliable, and therefore it will reduce the possibility of accurate planning.

The basic idea of the inventory-based approach is to see the QCs as customers, which need to be served with goods. In AGV systems without the usage of cassettes, the procedure of the inventory-based works like a waiting buffer in the area where the arriving AGVs wait in order to serve the QCs. The policy of the inventory-based approach builds upon inventory management where the goal is to have no customer waiting for goods, which means that the inventory level should never be zero. Another aspect of inventory management is that the inventory level should not be too high either, ideally the best scenario of this the inventory-based approach is that an AGV serves a QC with the lowest inventory level just in time.

According to the inventory-based approach, every QC has an inventory level. The inventory level consists of the waiting AGVs plus those that are on their way to serve that specific crane. Therefore, ILA_q is defined as the number of AGVs that are busy with current loading jobs of QC q, but have not yet reached q. Consequently, the number of waiting or in motion AGVs for the discharging crane (p) is defined as ILA_p. However, these definitions are not entirely suitable for this purpose, and another factor needs to be weighted in, namely the AGV travel and service time. The travel and service time consists of driving to the stacking area, waiting for service and finally driving back to a QC. Therefore, the inventory level of a discharging quay-crane (p) must be lower than a loading one (q) in order to reach the same productivity level. To solve the aspect researchers have introduced a parameter phase (ϕ) factor that is defining the inventory level of a loading quay-crane higher. So therefore the definition of $ILA'_q = \dfrac{ILA_q}{\phi}$, but the definition of the ila_p is equal to ila'_p. If a tie occurs between the two inventory levels ($ila'_q = ila'_p$) the quay-crane with the last visited AGV is more urgent. In addition to the reduction of the usage of due times, this method can be incorporated with other operational issues in practice, for example, the priority of the QCs can easily be implemented. QCs that serve high priority vessels will receive a reduction on their ila'_q in order to make the jobs for those cranes appear more urgent. Altogether, the straightforwardness of the inventory idea makes it quite interesting and highly applicable with respect to practical needs.

Looking more closely to the assignment of tasks, an AGV (a) that is free for a new job should be assigned to the first unassigned job of the QC (q) that has the lowest inventory (buffer). This makes sure that the most likely urgent QC receives an AGV first, which will reduce waiting times and shorten the AGV queues. In order to assign the n jobs to the n AGVs the inventory-based solution has created a standard linear assignment problem with a cost calculation that consists of three components:

- An AGV a may have a current job that must be completed before it can start the next empty travel. The estimated waiting time for availability w_a obviously influences the duration until the next job j can be started as well as the time duration until the AGV can arrive at the related quay-crane. Note that w_a is zero if AGV a does not have a current job.

- According to the pick-up location of job j and the current position of AGV a there is an expected empty travel time e_{ja} if j is assigned to a. This empty travel time affects the arrival of the AGV at the quay-crane.
- We define $1 \leq o_j \leq n$ as the ordinal number of job j according to the order in which the jobs were chosen for assignment. That is, job j with $o_j = 1$ is the most urgent job with respect to the inventory levels ila'_q, job i with $o_j = 2$ as the second most urgent job.

With this at hand the cost is defined as the following formula: $c_{ja} = (\lambda * (n - o_j) + 1) * (w_a + e_{ja})$. Using this formula to determine the costs, the authors have used the well-known Hungarian method to resolve the resulting assignment problem in order to minimize the total costs of assignments.

4.4 IMPLEMENTATION STUDY

Data collection relating to real factory equipment is now more accurate and timelier. In this past, this task took weeks, with many engineers needed to scan and measure the production facility. By eliminating that step, the new IoT technology will permit faster modifications, a critical requirement for autonomous production. We already know that manufacturing is a key driver for economic growth, attracting investment, spurring innovation, and creating high-value jobs. All of the breakthrough and developments that are happening now – the building blocks for autonomous production – are already adding economic value. Imagine what will happen when autonomous production is routine. We also see autonomous production as a way to address many global challenges such as a growing and aging population, climate change and resource scarcity.

To verify the proposed mathematical model of path planning and production optimization an illustrative example is shown and solved using GAMS 24.1.2 optimization software. The number of jobs is 2, number of cells is 4, number of vehicles is 1 and all available paths are 3. The processing times and waiting times for the jobs in any cells are:

$$P_{ir} = \begin{bmatrix} 10 & 7 & 9 & 13 \\ 11 & 8 & 14 & 12 \end{bmatrix}; \quad WT_{ir} = \begin{bmatrix} 1 & 2 & 1 & 3 \\ 2 & 3 & 1 & 2 \end{bmatrix},$$

Distance time between any two cells is:

$$D_{er} = \begin{bmatrix} 1000 & 10 & 1000 & 14 \\ 10 & 1000 & 12 & 17 \\ 1000 & 12 & 1000 & 8 \\ 14 & 17 & 8 & 1000 \end{bmatrix}$$

Based on this input information, the proposed approach has been applied for each of the considered scenario nodes. The results of the whole approach applied on scenario path $o_0 - o_{1A} - o_{2B}$ are reported in Table 4.1.

TABLE 4.1
Numerical results for the industrial real case

	Cost type	t_1	t_2	t_3	Total
Robust approach	Robot purchase	358,883	0	0	358,883
(overall approach)	Module purchase	50,000	0	0	50,000
	Reconfiguration	0	0	0	–
	Storage	0	12,000	0	12.000
	Operative	92,010	106,002	78,894	276,906
	Tool purchase	45,000	20,000	20,000	85,000
	Total (discount)	545,893	133,412	92,425	771,730
Single path	Robot purchase	358,883	0	0	358,883
optimum	Module purchase	40,000	0	10,000	50,000
(best configuration	Reconfiguration	0	10,000	10,000	20,000
is chosen for	Storage	0	18,000	0	18,000
each scenario)	Operative	100,776	103,542	83,850	288,168
	Tool purchase	45,000	20,000	20,000	85,000
	Total (discount)	544,659	146,502	115,748	806,909
Single node	Robot purchase	358,883	0	Infeasible solution	358,883
optimum	Module purchase	40,000	0	Infeasible solution	40,000
(best o_0	Reconfiguration	0	10,000	Infeasible solution	10,000
configuration	Storage	0	18,000	Infeasible solution	18,000
is used in every	Operative	100,776	103,542	Infeasible solution	204,318
scenario)	Tool purchase	45,000	20,000	Infeasible solution	65,000
	Total (discount)	544,659	146,502	–	691,161

The first row refers to the robust solution. The second row refers to the optimal solution for the considered scenario path only, obtained by choosing the best configuration solution at each step. The last row reports the solution in which the optimal solution for o_0 is used in every time bucket. The solutions are compared in terms of purchasing, reconfiguration, storage and operational costs. Results demonstrate that the robust solution ensures a lower total discounted cost compared to the optimal solution for the single scenario path ($771 730 against $806 909), the difference is mainly due to the fact that the robust solution behaves proactively, purchasing additional pieces of equipment in advance, while the other solution has to react to the changes through a reconfiguration step, whose impact on the cost is relevant ($10 000).

As stated before, the model minimizes the total delay time of AGVs. In this example the number of production cells is considered to be 4. All jobs have certain processing and waiting times as inputs to the model. In the process of numerical optimization, instead of the nonexistent path a large value (in this example, 1000) is set for the distance time parameter.

The output of the optimization is given below:

Objective value: Z = 43,

Optimization time = negligible,

The assignments for the decision variable $Y_{vierm} = 1$, are

$$Y(1,1,1,4,1) = 1 \quad Y(1,1,2,1,1) = 1$$

$$Y(1,2,1,4,1) = 1 \quad Y(1,2,2,6,1) = 1$$

$$Y(1,2,3,2,1) = 1 \quad Y(1,2,4,3,1) = 1$$

An instance for the obtained assignment is $Y(1, 2, 1, 4, 1) = 1$, that is, Vehicle 1 handles Job 1 from Cell 3 to Cell 2 in Path 1.

4.5 DISCUSSION

In order to implement the agents and entities presented in the simulation model, the object-oriented programming language Java has been used together with packages from the open-source simulation framework DESMO-J (Discrete-Event Simulation and Modelling in Java). DESMO-J has been developed by the University of Hamburg, and is targeted at discrete event modeling and simulation with additional white box functionality that enables the modeler to create model-specific entities with active behavior. Due to the variety of features that DESMO-J offers, a process-oriented implementation was selected mainly because processes within DESMO-J are "conceptually active entities which package both properties and behavior". Every active entity or process within DESMO-J has its own life-cycle which describes its behavior. Data structures that are included in the process are declared as properties.

The simulator has two main abstract classes that extend SimProcess from DESMO-J, namely crane and *AGV*. These two classes include the basic data structures and methods (activities) that are needed for its children (the different types of *AGVs*, (Stack Crane (*SC)* and *QC)*. YardController which extends the Model from the DESMO-J works like a singleton, and is responsible for initializing the simulation with creating the different active and non-active entities that are included in the simulation. It is also responsible for the communication between the different entities during the simulation, which also includes dispatching of the *AGVs*.

If we start with the *QC* it will check its buffer for any empty cassettes, and ask the YardController for the next job. If the *QC* receives a job and has an empty cassette, it will start unload the job from the ship. This activity is implemented in the simulator with the hold function that DEMSO-J provides. For how long a *QC* holds is determined by the job specification and parameters such as *QC* throughput and interruption, which are set by the user before a simulation. However, one time unit in the simulator represents one second in real time. When the unloading of the job is done, it will notify the YardController that the job is done, which will depending on the dispatching strategy that has been assigned to an *AGV*. YardController will also check whether the *QC* needs more cassettes, and if so creates a returning cassette job to that specific *QC*. The *QC* will continue unloading jobs from the ship until it has no more empty cassettes, and then passivates which is a undetermined hold state where the *QC* waits until it receives a notification from the YardController that a new empty cassette has been delivered. This life cycle will continue until all the jobs have been unloaded from the ship. If we take a closer look at the *SC* which have a very similar lifecycle as the *QC*, but with the difference of that it only stacks containers when it receives notification from the YardController that a job has been delivered by an *AGV*, and is ready for stacking.

The third and last active entity is *AGV*, but dependent on dispatching strategy it differs in its life cycle, which we will discuss more in further sections in this chapter. Other entities that are included in the simulator are cassette and job. These two are typically non-active entities with no lifecycles, and with the purpose of storing and receiving data. The simulator has a few more classes such as a database class and user interface class, but these classes has no importance when it comes to the core functionality of the simulator, and therefore no further description will be given.

4.5.1 THE GREEDY ALGORITHM

As mentioned above, the simulator has different types of *AGVs* depending on the dispatching strategy. One of these types is the *GreedyAGV*, which has incorporated a simple Greedy-based strategy for dispatching. The *AGV* strictly follows its job list with a FIFO (First In First Out) strategy, and does not consider any due times or any priority when selecting a job. The *GreedyAGV* is also responsible for calculating the cost for a future job already adding the existing ones in order to give an estimate of when future jobs can be done

The assignment of jobs is quite straightforward, the YardController asks all of the *GreedyAGVs* for a cost estimation when a job can be completed. As mentioned above, the *GreedyAGV* will base its estimation on cost estimations of all the jobs in its job list plus the new one. Based on those estimations the YardController will strictly assign jobs to the *AGV* with the lowest cost estimation of completion time for that specific job. Another consideration that is worth mentioning is that a job with the Greedy approach is only dispatched when it is ready for dispatching. Therefore, if there are no jobs at a certain time a *GreedyAGV* will simple wait at the current location until it receives a job from the YardController.

4.5.2 THE GREEDY WITH LOOK-AHEAD ALGORITHM

The basic concept of the Greedy Look-Ahead algorithm is to improve the simplistic heuristic that the normal Greedy Algorithm uses, with the intention of using combinations of jobs in order to select which *AGVs* are the best for a set of specific jobs. During the implementation of this algorithm, we discovered a couple of aspects that lead to serious problems, which lead to two versions of the Look-Ahead dispatching algorithm. The main problem with implementing this algorithm in our view is the fact that the usage of cassettes brings a lot more complexity to the domain compared with old *AGV* systems without cassettes. Due to the existence of return cassette jobs which will be created during the simulation with specifications that are dependent on variables such as when another job is done, how many cassettes are available at a certain time and so on.

Another factor that needs to be mentioned is that with the introduction of interruption, combined that with the complexity of adding the cassette feature, it is impossible to create a combination of future jobs that is accurate, because without good estimations on service times on jobs, return cassette jobs will be created at uncertain points with uncertain specifications. For instance if you have two *QCs* in the simulation, and at one *SC* you have an empty cassette that is ready to be placed in any *QC* buffer. Which *QC* gets that empty cassette is uncertain if you cannot accurately estimate which of them is gets an empty position in the buffer first. Obviously, one solution to this problem might be to assign cassettes to a specific *QC*, and when a cassette is empty it should always return to its assigned *QC* buffer. However, in this simulator the cassettes can be used by any *QC* buffer and are not assigned to a specific *QC*. It is quite obvious that this approach gives more flexibility in the handling of cassettes.

During the implementation, we implemented two alternatives to solve the problem with unknown return cassette jobs without compromising the use of combinations. Both our solutions used similar strategies to the Greedy, but in one of the alternatives, dispatching also occurred on unready jobs, which gave the *LookAGVs* the capability of driving to a crane before a job is ready. However, this method excluded the unknown return cassette jobs when it was creating combinations of future jobs. The second solution that was implemented was a similar dispatching approach as described for the *GreedyAGV*, but with the difference of adding future *QC* jobs that are certain together with the job that is up for dispatching, and creating the different combinations. From the best combination from a time standpoint, the *AGV* that was assigned to that specific job was dispatched. The advantage to this solution was that jobs were dispatched in relatively updated states, and the dispatching strategy was not so sensitive to delays, but the downside was that the *AGVs* were not predetermined for jobs, and therefore did not have the ability to to drive to a crane in advance of a job being ready. The results from both our alternatives were somewhat unconvincing which we will elaborate more in future chapters.

With the life-cycles from our two alternatives, the second one is very similar to the normal *GreedyAGVs*, and the first one is somewhat closer linked to the *InvAGV* that are presented in the next section. The calculation and creation of combinations, which is done in YardController, can be viewed in the code snippet in Appendix A. Both alternatives use the same algorithm, with the difference that in the second alternative just the job that is up for dispatching is actually dispatched.

Looking more closely at the Look-Ahead algorithm shown above, the first thing that happens is that the jobs are sorted by estimated due times. When that is taken care of, and when the right number of jobs have been selected based upon the looking forward factor meaning how many jobs that the algorithm will look ahead at, the YardController will then ask all the *LookAGVs* for an initial cost indication when free and ready to take a new job. When that has been finalized a Variator (class that creates combinations) will construct every possible dispatching combination for those jobs, and the YardController will calculate based upon the *AGV* estimation calculation the cost of every combination, and compare it with the goal of finding the best combination with the lowest time cost. The final thing that happens in the algorithm is that the *AGVs* are dispatched or in alternative two, an *AGV* is dispatched for the specific job.

4.5.3   THE INVENTORY ALGORITHM

The Inventory Algorithm is the only one in this thesis that is not using any due times or job estimations to determine which jobs that are to be assigned to which *AGVs*. The main purpose of the Inventory Algorithm is to always be located at the most needed crane meaning the crane with the highest inventory level. The inventory approach described earlier in this chapter is basing their inventory levels on the last *AGV* service time. Because this method was created for *AGV* systems without a cassette, we have based our inventory levels upon the number of cassettes in the buffer.

Looking at the lifecycle of the *InvAGV*, the procedure is to ask the YardController which crane has the highest inventory level based upon what its current position is. The InvAGV will then drive to that returned crane and check whether there are any jobs at that crane at this moment. If there is no job the *InvAGV* will simply wait for an undetermined period, and is only reactivated when a new job is created from that crane.

As mentioned above the inventory levels are based upon the number of cassettes located in the crane's buffer. However, looking at the algorithm, which is in Appendix A, a few more things are added to the algorithm in order to make it work in a sufficient way. Firstly, the number of *AGVs* at one crane will never exceed the number of cassettes in the buffer. Secondly, if a job is already ready for dispatching there is no need for InvAGV to determine all of the cranes' inventory levels, it will simply drive to the crane that initiated the first job in the job list. However, the dispatching strategy do not consider any job estimations of the *AGV*, and it simply assigns jobs on a first come first served basis.

4.5.4   HYBRID BETWEEN GREEDY AND INVENTORY

During the implementation of the different dispatching algorithms, an idea of a new dispatching strategy emerged. The idea was to combine the Greedy-based algorithm with the inventory, which will give the capability to the *AGV* to both estimate the cost of a future job, and still consider inventory levels before a job is ready for dispatching.

Because of the similarities between the previously mentioned algorithms, the lifecycle and the algorithm will not be presented. In addition to that, no model has been created for it. However, it should be mentioned that because of the time constraints, this algorithm was not completely tested in order to optimize it, and therefore could it be interesting to further study this concept before it is completely discarded.

Typical issues of the discrete manufacturing business include the complexity of the product mix, small batch sizes, variety of treatments per component and last but not least the cyclical fluctuations in market demand. Transport of such products requires a flexible logistic solution. AGV systems provide a maximum of free space on the shop floor and are easily adoptable to changes in the plant layout and process equipment. The assignments obtained from the proposed mathematical model for the path planning problem in the manufacturing system emphasizes the guide paths and AGVs being applied in the routing and scheduling. This information in an industrial system helps the decision makers to identify the effective elements on the system for performance evaluation and productivity analysis. Another aspect of the model is to find out the most visited paths and locate the maintenance department around them while designing or redesigning the shop floor, to reduce the repair or maintenance activity times.

4.6   CONCLUSIONS

This chapter has developed a mathematical model to optimize optimal path planning for AGVs in a manufacturing system. The IoT needs both data collection and actuation features but it also needs contextual information and orchestration to make this data useful and enable automation scenarios to be built around it. With the emerging information technologies, such as the IoT, big data, and cloud computing together with artificial intelligence technologies, we believe the smart factory of Industry 4.0 can be implemented. To analyze the capability and effectiveness of the model an example is solved. Important drivers for the manufacturing and discrete manufacturing industries are delivering products on time while improving quality and reducing production costs. AGV systems comprise of intelligent scheduling software to meet just-in time deliveries and to save costs related to moving products internally. As they manoeuvre very carefully and handle goods with care damage to products and equipment is minimized.

BIBLIOGRAPHY

Berman, S., Edan, Y., & Jamshidi, M. (2003). Decentralized autonomous automatic guided vehicles in material handling. *IEEE Transactions on Robotics and Automation*, 19(4), 743–749.

Delorme, X., Dolgui, A., & Kovalyov, M.Y. (2012). Combinatorial design of a minimum cost transfer line. *Omega*, 40(1), 31–41.

Fazlollahtabar, H. (2016). Parallel autonomous guided vehicle assembly line for a semi-continuous manufacturing system. *Assembly Automation*, 36(3), 262–273.

Fazlollahtabar, H. (2018a). Lagrangian relaxation method for optimizing delay of multiple autonomous guided vehicles. *Transportation Letters*, 10(6), 354–360.

Fazlollahtabar, H. (2018b). Scheduling of multiple autonomous guided vehicles for an assembly line using minimum cost network flow. *Journal of Optimization in Industrial Engineering*, 11(1), 185–193.

Fazlollahtabar, H. (2019a). An effective mathematical programming model for production of automatic robot path planning. *The Open Transportation Journal*, 13(1), 11–16.

Fazlollahtabar, H. (2019b). Triple state reliability measurement for a complex autonomous robot system based on extended triangular distribution. *Measurement*, 139, 122–126.

Fazlollahtabar, H. (2020). Comparative simulation study for configuring turning point in multiple robot path planning: Robust data envelopment analysis. *Robotica*, 38(5), 925–939.

Fazlollahtabar, H. (2021). Robotic Manufacturing Systems Using Internet of Things: New Era of Facing Pandemics. *Automation, Robotics & Communications for Industry*, 4, 82.

Fazlollahtabar, H. (2022). Internet of Things-based SCADA system for configuring/reconfiguring an autonomous assembly process. *Robotica*, 40(3), 672–689.

Fazlollahtabar, H., Es'haghzadeh, A., Hajmohammadi, H. et al, (2012). A Monte Carlo simulation to estimate TAGV production time in a stochastic flexible automated manufacturing system: a case study. *International Journal of Industrial and Systems Engineering*, 12(3), 243–258.

Fazlollahtabar, H., & Hassanli, S. (2018). Hybrid cost and time path planning for multiple autonomous guided vehicles. *Applied Intelligence*, 48, 482–498.

Fazlollahtabar, H., & Jalali, S. G. (2013). Adapted Markovian model to control reliability assessment in multiple AGV. *Scientia Iranica*, 20(6), 2224–2237.

Fazlollahtabar, H., & Mahdavi-Amiri, N. (2013a). Design of a neuro-fuzzy–regression expert system to estimate cost in a flexible job shop automated manufacturing system. *The International Journal of Advanced Manufacturing Technology*, 67(5/8), 1809–1823.

Fazlollahtabar, H., & Mahdavi-Amiri, N. (2013b). Producer's behavior analysis in an uncertain bicriteria AGV-based flexible job shop manufacturing system with expert system. *The International Journal of Advanced Manufacturing Technology*, 65(9/12), 1605–1618.

Fazlollahtabar, H., & Mahdavi-Amiri, N. (2013c). An optimal path in a bi-criteria AGV-based flexible job shop manufacturing system having uncertain parameters. *International Journal of Industrial and Systems Engineering*, 13(1), 27–55.

Fazlollahtabar, H., & Niaki, S.T.A. (2017a). Binary state reliability computation for a complex system based on extended Bernoulli trials: Multiple autonomous robots. *Quality and Reliability Engineering International*, 33(8), 1709–1718.

Fazlollahtabar, H., & Niaki, S.T.A. (2017b). Integration of fault tree analysis, reliability block diagram and hazard decision tree for industrial robot reliability evaluation. *Industrial Robot: An International Journal*, 44(6), 754–764.

Fazlollahtabar, H., & Niaki, S.T.A. (2017c). *Reliability Models of Complex Systems for Robots and Automation*. CRC Press.

Fazlollahtabar, H., & Niaki, S.T.A. (2018a). Cold standby renewal process integrated with environmental factor effects for reliability evaluation of multiple autonomous robot system. *International Journal of Quality & Reliability Management*, 35(10), 2450–2464.

Fazlollahtabar, H., & Niaki, S.T.A. (2018b). Modified branching process for the reliability analysis of complex systems: Multiple-robot systems. *Communications in Statistics-Theory and Methods*, 47(7), 1641–1652.

Fazlollahtabar, H., Mahdavi-Amiri, N., & Muhammadzadeh, A. (2015). A genetic optimization algorithm for nonlinear stochastic programs in an automated manufacturing system. *Journal of Intelligent & Fuzzy Systems*, 28(3), 1461–1475.

Fazlollahtabar, H., & Olya, M.H. (2013). A cross-entropy heuristic statistical modeling for determining total stochastic material handling time. *The International Journal of Advanced Manufacturing Technology*, 67(5/8), 1631–1641.

Fazlollahtabar, H., Rezaie, B., & Kalantari, H. (2010). Mathematical programming approach to optimize material flow in an AGV-based flexible job shop manufacturing system with performance analysis. *The International Journal of Advanced Manufacturing Technology*, 51(9–12), 1149–1158.

Fazlollahtabar, H., & Saidi-Mehrabad, M. (2015a). *Autonomous Guided Vehicles: Methods and Models for Optimal Path Planning*. Germany: Springer International Publishing.

Fazlollahtabar, H., & Saidi-Mehrabad, M. (2015b). *Autonomous Guided Vehicles: Methods and Models for Optimal Path Planning*. Springer International Publishing, Switzerland. ISBN 978-3-319-14746-8

Fazlollahtabar, H., & Saidi-Mehrabad, M. (2015c). Risk assessment for multiple automated guided vehicle manufacturing network. *Robotics and Autonomous Systems*, 74, 175–183.

Fazlollahtabar, H., & Saidi-Mehrabad, M. (2019). *Cost Engineering and Pricing in Autonomous Manufacturing Systems*. Emerald Publishing Limited.

Fazlollahtabar, H., Saidi-Mehrabad, M., & Balakrishnan, J. (2015a). Mathematical optimization for earliness/tardiness minimization in a multiple automated guided vehicle manufacturing system via integrated heuristic algorithms. *Robotics and Autonomous Systems*, 72, 131–138.

Fazlollahtabar, H., Saidi-Mehrabad, M., & Balakrishnan, J. (2015b). Integrated Markov-neural reliability computation method: A case for multiple automated guided vehicle system. *Reliability Engineering & System Safety*, 135, 34–44.

Fazlollahtabar, H., Saidi-Mehrabad, M., & Masehian, E. (2015). Mathematical model for deadlock resolution in multiple AGV scheduling and routing network: a case study. *Industrial Robot: An International Journal*, 42(3), 252–263.

Fazlollahtabar, H., Saidi-Mehrabad, M., & Masehian, E. (2021). Robotic industrial automation simulation-optimization for resolving conflict and deadlock. *Assembly Automation*, 41(4), 477–485.

Fazlollahtabar, H., & Shafieian, S. H. (2014). An optimal path in an AGV-based manufacturing system with intelligent agents. *Journal for Manufacturing Science and Production*, 14(2), 87–102.

Herrero-Pérez, D., & Martínez-Barberá, H. (2010). Modeling distributed transportation systems composed of flexible automated guided vehicles in flexible manufacturing systems. *IEEE Transactions on Industrial Informatics*, 6(2), 166–180.

Khan, M., Hussain, I., & Noor, S. (2011). A knowledge based methodology for planning and designing of a flexible manufacturing system (FMS). *International Journal of Advanced Manufacturing Systems*, 13(1), 95–109.

Michalos, G., Makris, S., & Mourtzis, D. (2012). An intelligent search algorithm-based method to derive assembly line design alternatives. *International Journal of Computer Integrated Manufacturing*, 25(3), 211–29.

Mourtzis, D., Vlachou, E., & Milas, N. (2016). Industrial big data as a result of IoT adoption in manufacturing. *Procedia CIRP*, 55, 290–295.

Ravi Raju, K., & Krishnaiah Chetty, O.V. (1993). Addressing design and control issues of AGV-based FMSs with Petri net aided simulation, *Computer Integrated Manufacturing Systems*, 6(2), 125–134.

Rymaszewska, A., Helo, P., & Gunasekaran, A. (2017). IoT powered servitization of manufacturing–an exploratory case study. *International Journal of Production Economics*, 192, 92–105.

Shojaeifar, A., Fazlollahtabar, H., & Mahdavi, I. (2016). Decomposition versus minimal path and cuts methods for reliability evaluation of an advanced robotic production system. *Journal of Automation Mobile Robotics and Intelligent Systems*, 10(3), 52–57.

Thramboulidis, K., & Christoulakis, F. (2016). UML4IoT—A UML-based approach to exploit IoT in cyber-physical manufacturing systems. *Computers in Industry*, 82, 259–272.

Xidias, E., & Azariadis, P. (2011). mission design for a group of autonomous guided vehicles. *Robotics and Autonomous Systems*, 59, 34–43.

5 Industrial IoT and Sustainability

5.1 INTRODUCTION

In the recent era, investigators have been faced with important challenges in terms of technological revolutions. Indeed, intelligent machines and robots, Cyber-Physical Systems (CPS), the Internet of Things (IoT), Big Data, virtual industrialization and Smart Factories are creating new possible futures in terms of society (Blanchet, Rinn, Von Thaden, & De Thieulloy, 2014; Hermann, Pentek, & Otto, 2016). The fourth industrial revolution with the general introduction of a set of new devices, processes, and technological tools was streamed but not well integrated (Blanchet et al., 2014; Wahlster, 2012). By the evolution of Industry 4.0 and the related technologies, the logistics systems were smarter and more flexible in their ability to adopt real time data and to make the corresponding decisions. By the way, coordination among all elements of the logistics system should be good enough to be able to handle such a dynamic environment. Thus, we aim to integrate the process and categorize the factors and functions of each element in any stage. Also, to relax the challenges of data flow and deposit, the Internet of Things, big data, and the cloud environment are employed.

The remainder of this chapter is organized as follows. Next, the state of the art is reviewed. In Section 3, the information flow for a logistics system is developed. In Section 4, the proposed integrated model on the platform of the Internet of Things is illustrated and the required functions are reported. We conclude in Section 5.

5.2 RELATED WORKS

The keyword "Industry 4.0" means a development that essentially made changes to traditional industries (Rennung, Luminosu, & Draghici, 2016). As far as the field of logistics is concerned, major implications are predicted, too. In fact, logistics represents an appropriate application area for Industry 4.0 (Bauernhansl, Ten Hompel, & Vogel-Heuser, 2014). The integration of CPS and the IoT into logistics promises to enable a real-time tracking of material streams, improved transport forwarding and accurate risk management, to mention but a few prospects. In fact, it can be argued that Industry 4.0 in its pure perspective can only become reality if the logistics is able

DOI: 10.1201/9781003400585-5

to provide production systems with input factors needed at the right time, of the right quality and in the right place.

In addition, industrial companies are facing new challenges in terms of increasing individualization of products, the need to increase resource effectiveness and reducing marketing time. These problems are faced in particular with increasing digitization, IT penetration and networking of products, manufacturing resources and processes. The concepts are often classified under the name "Industry 4.0" (Lachenmaier, Lasi, & Kemper, 2015). According to the services industry, the dominant logic, with the focus on adapting traditional engineering approaches to the service sector, leads to the emergence of an official approach to service engineering (Qiu, 2014), In this term, some authors call it a "servitization" process (Baines, 2015; Schmenner, 2009).

Virtual Manufacturing (VM) introduced the concept of manufacturing process simulation using special computing environments, but the production plan was still connected to the product, which added a business model (van der Aalst & Arthur, 2003). The recent models of the Internet of Industry (Industry 4.0) can ultimately integrate business process and production, and fully express the concept of the product/ service, which is based on physical-cyber production systems (Brettel, Friederichsen, Keller, & Rosenberg, 2014; Hermann et al., 2016).

The IoT offers new possibilities for performance. For example, road transport trucks can be automatically controlled by the host's profile, which will allow them to work at predefined intervals at standard speeds to increase fuel economy. The Daimler Group has invested in the development of mobile services such as car2go, myTaxi or moovel; General Electric has also invested in systems for launching and operating equipment, and factories have been using the "Industrial Design" system (the Internet industry) use (Perera, Ranjan, Wang, Khan, & Zomaya, 2015).

Using the IoT, it is possible to monitor the travel process of packages and letters. Continuous monitoring and control create the possibility of solving the question, "Where is the package?" In case of delay, the client can be notified before there are important consequences. In the case of storage in warehouses, intelligent shelving and smart pallets will become the driving force behind modern financial management. In terms of freight transportation, tracking is quicker, more accurate, predictable and secure. Analysis of the development of a "connected fleet" can help predict failure and automatically plan currents aimed at improving the supply chain.

Also, big data has a potential value that has not yet been discovered. Nevertheless, it has been observed that "Every manufacturer has an unbelievable amount of data that is never put to use. They are literally drowning in it, and when they begin to gather it, analyze it and tie it to business outcomes, they are amazed by what comes out," " (Records & Fisher, 2014). Several researchers worked on the applications of big data. Chen et al. (2012) developed the concept of big data in business intelligence. Dubey et al. (2016) investigated sustainability in manufacturing using big data. Song et al. (2016) studied performance evaluation in a big data environment. Wamba and Akter (2015) analyzed the literature for big data in supply chain mangement. Wang et al. (2016) studied the logistics system for research and applications for big data. Generally, researchers tried to study different dimensions of big data and gain potential benefits for Supply Chain Management (SCM). Understanding the role of big data in increasing the efficiency and profitability of a company to supply chain managers is important.

5.3 INFORMATION FLOW OF LOGISTICS SYSTEMS

Supply chain design is a related decision process that aims to define appropriate configuration and efficient management strategies for these systems. In addition, in order to help understand the logistics system's operation and how it is structured and organized in a hierarchy, and the ability to balance and plan, several aspects of this process, such as nutrition, equipment selection, ergonomic risk and learning effects, need to be considered.

For a generic logistics system including multiple layers such as supplier, producer, distributor, retailer and customer the impacts of the internet can be considered. The information flow between any two layers of the logistics system influences decision making. Decisions are significant for improving performance and increasing productivity.

We have the logistics system information flow that includes ordinary logistics parts. In the first stage we have planning which means that before any action we should prepare a plan in order to optimize our logistics stages. After that we have sourcing, which represents the decisions about suppliers and materials to reduce the effects of material costs. In the making stage, suppliers and material costs will be considered to specify manufacturing costs. In the delivery stage, warehouses and distribution centers will specify inventory costs which will influence customers in the buying stage.

As shown in the proposed logistics system, we consider five main stages where each stage contains multiple choices due to multiple factors being considered. These factors show that in each stage several partners exist, that is, in the sourcing stage several enterprises are active in material provision. In addition, between all of the system stages the information exchange is shown by arrows. This information exchange between each stage happens continuously, between the sourcing and the making stage the information relates to suppliers delivering to manufactures. Similarly, between the making stage and the delivery stage the information relates to manufacturers delivering to distribution centers. Similarly between the delivery stage and the buying stage the information relates to distribution centers delivering to costumers. The type of information that is exchanged between stages will be different, for example, between the making and delivery stage, information such as production type, number of products and quality measures will be delivered to distribution centers.

For all material flows in a logistics system we consider transportation costs and their implications in order to obtain optimal solutions effectively. Also, we recognize that in most of the actions and reactions of this logistics flow we can use the IoT-based information systems. On this subject, we should note that the IoT-based information flow system can be used in most of the actions and reactions of this logistics stream. In most companies in the logistics and transportation sectors, there is a need to use internet-based termination of objects. In fact, the internet will improve the supply chain objects. The IoT will help companies increase their safety and impact levels.

Key technologies in the implementation of the internet are objects of Wi-Fi connectivity, security sensors, and Near Field Communications (NFCs). One of the concerns is privacy and information security, which could be seen as the biggest obstacle in implementing IoT solutions. Also, the high level of complexity of these solutions is important as a high performance risk.

5.4 INTEGRATED IOT-LOGISTICS SYSTEM

In this research, an intelligent logistics information flow system is designed based on the Internet of Things (IoT). In fact, elements of the logistics system are considered as things having data exchange for operations management and optimization purposes. As shown in Figure 5.1, operations and decisions in a logistics system are categorized as several items to be processed using data extracted from its elements. Operations are divided into production, transportation, inventory and distribution. Each has some sub-operations. For instance, production is composed of machines and equipment that are interconnected using Wi-Fi technology to exchange data. Also, production processes are monitored in an online web system. On the other hand, decisions are based on optimization performed using operational data. The most used decisions in supply chains are production planning, inventory control, vehicle routing, cost analysis and time management. Intelligent data transfer among operation and decisions elements leads to effective decisions.

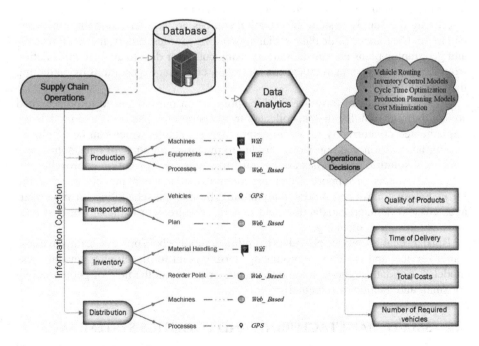

FIGURE 5.1 An IoT-based logistics system.

As a complex integrated system, the IoT complements various devices with sensor capabilities, identification, processing, communications, and network integration. An IoT system includes Industrial Wireless Networks (IWN) and the Internet of Things (IoT). This includes automation tools, networks, cloud computing and terminals. An IoT system is created by providing specific and personalized products. Users can customize products through web pages. Then, web servers direct information through wired or wireless networks to clouds and industrial plants. In order to get information, the designer integrates the design, and optimizes, manages and monitors the production process based on the products being manufactured effectively. With the help of self-optimizing decision making machinery, both machinery and equipment to improve performance will be further developed. Since the production and supply are dynamic, the life cycle of a product is also variable. With regard to changes, decentralization, self-optimization, and automation, they can contribute more effectively to the dynamic process.

Technologies that are employed in the proposed IoT-based supply chain are Wi-Fi for data exchange in production operation, the web service for inventory data collection and GPS that is used in vehicle routing for location determination purposes. A configuration of technologies and the corresponding operations of the supply chain are presented.

With respect to the proposed configuration of SCM, that could be very complex when multiple products in multi periods exist. To handle such a huge amount of information, specific paradigms and concepts need to be considered. To handle this, the idea of big data and various models are investigated.

Among the supply system, information sources are retailers, shipping, invoices, and more. Customer profile data, social networking profiles, orders, market forecasts, and geographic plans are substantial. By using customer data to analyze information from the delivery system, retailers can meet their customer expectations by predicting their behavior.

Thus, mathematical formulations for optimization purposes need to be designed so that they can handle big data collected from operations. The presented models are implemented concurrently to obtain comprehensive results which can be employed in optimal decision making. However, due to the larger amounts of data, sometimes numerical solution approaches are not effective. Intelligent solution approaches are also used as decision supports. The most common mathematical models are lot sizing for inventory control and production planning, production scheduling for time management and demand satisfaction, and material requirement planning for cost analysis and inventory planning.

Big data enables service providers to optimize supply chain processes, improve customer service, and presents "a promising starting point for developing new business models". Big data suggests some instruments operating in the field of geomarketing for small and medium-sized enterprises.

5.5   SMART MANUFACTURING AND LOGISTICS SYSTEM: SMLS

In order to consider an SMLS, an integrated manufacturing and logistics system based on the Internet of Things is designed. For each process of the system, the stages of preparation, supply, production, packaging, transportation, inspection, monitoring and evaluation, sales, and the like, are considered. Due to the cyber physical hardware available in the integrated SMLS such as Wi-Fi, Bluetooth, cloud computing and the web, the Global Navigation System (GPS), smart robots, electrical sensors and the like, the structure of system information exchange is implemented and the design steps are described.

Configuration of a SMLS in a multi-product and multi period environment could be very complex. To handle such a large amount of information, special paradigms and concepts need to be considered. To address this, the concept of big data is investigated. To make use of big data in the process of decision making, useful techniques are required. Therefore, mathematical formulation is employed to optimize the information collected from big data in operations management, which will simultaneously be implemented to obtain the comprehensive results used in optimal decision making. Hence, in some cases with large volume of data, numerical solution methods are not effective enough. Smart solution methods are also used as decision support. The most common mathematical models in this case are inventory measurement for inventory control and production planning, scheduling for time management and demand satisfaction, and planning materials for cost analysis and financial planning.

In order to extract data from the SMLS structure of information flow in the decision making process, it is necessary to design mathematical models in different stages of manufacturing and logistics tasks. For this purpose, we need to perform multistage optimization using the mathematical models for each step.

In this research, we present three models for resource supply, product generation and distribution in the general mode. The resource supply model is to minimize the cost of purchasing raw materials from suppliers and transfer them to the inventory, the production planning model is to minimize production process and inventory costs, and the distribution model is to minimize the costs of distribution and delivery of the final products to customers.

In the following, we refer to the general mathematical models, and then we define the parameters and variables of each model, and then we will describe the model and its explanations.

5.5.1 MODELING ELEMENTS

Modeling elements for resource planning, production and distribution and deliverables models are:

- The volume of mass production or production categories
- Planning horizons
- Resources (can be saved, not saved)
- Cost of production
- Warehousing costs
- Customer service level
- Priority restrictions
- Loss of order

- Launch time, change time
- A set of items
- Short horizons
- For each item: unlimited ordering capacity
- Production planning related to capacity constraints
- Simultaneous production and purchase of all items
- From raw materials to final products
- Fulfillment of external demand
- Meet domestic demand
- Short horizons
- (The product is demanded)
- Material flow series, montage and general structure
- There is a possibility to deliver pre-term product demand.

5.5.2 General Assumptions

General assumptions are:

- All data is already available (demand forecast)
- All information is definitive (not accidental)
- The production chain structure is given.

5.5.3 Models

Considering the points raised and the structure of the information flow of logistic and manufacturing processes, three mathematical models for supply, production and distribution are presented below.

5.5.3.1 Model 1: Determine the order size for the procurement and procurement phase

5.5.3.1.1 Definitions

- $t = 1,\ldots,n$ Planning period
- $i = 1,\ldots,m$ Type of material
- d_{ti} the amount of substance i used in period t
- Purchase cost: fixed cost q + margin cost p
- p_{ti} is the unit purchase cost for the primary substance i in period t
- h_{ti} warehousing costs for raw material i during period t.

5.5.3.1.2 Decision variables

- x_{ti} volume of the purchase of material i in period t
- y_{ti} binary variable; the status of the purchase process for matter i in period t
- s_{ti} stock of the initial material i at the end of t period in stock.

5.5.3.1.3 *Formulation of the model*

$$\min \sum_{i=1}^{m} \sum_{t=1}^{n} p_{ti} x_{ti} + q_{ti} y_{ti} + h_{ti} s_{ti}. \tag{1}$$

such that:

$$s_{(t-1)i} + x_{ti} = d_{ti} + s_{ti}, \; \forall t, i, \tag{2}$$

$$s_{(0)i} = s_{(n)i} = 0, \; \forall i, \tag{3}$$

$$x_{ti} \leq M_{ti} y_{ti}, \; \forall t, i, \tag{4}$$

$$x_{ti} \in \mathbb{R}_+, s_{ti} \in \mathbb{R}_+, y_{ti} \in \{0,1\}. \tag{5}$$

Equation (1) represents the objective function of minimizing the final cost of purchasing raw materials from suppliers, considering unit purchase costs, fixed purchases and stockpile for all raw materials during the planning period.

The constraint (2) indicates that for each raw material supplied, the total inventory amount of that substance per period and the amount of material used during that period is equal to the total amount of inventory of that material in the previous period and the volume of the purchase of that product in that period. According to constraint (3), the model assumes that the inventory of raw materials in the warehouse before the beginning of the first period and after the end of the planning period is equal to zero, that is, the warehouse before the start of the planning period and after it is empty of raw materials. Limit (4) shows that for each raw material, the volume of purchases during the period when the purchasing organization ceases to be zero, and when the status of the buying organization is active the volume of purchase may increase to an unlimited amount. Finally, constraint (5) represents the type of decision variables, so that the volume of inventory and inventory for each raw material in each period can be a positive real number. How the purchasing organization works for each raw material in its collection or inactivity it is determined by a binary variable. The number 1 indicates the active purchase process for each raw material in the period and the number zero indicates inactivity of the buying process for that substance.

5.5.3.2 Model 2: Planning for the production stage

5.5.3.2.1 *Definitions*

- $t = 1, \ldots, n$ planning period
- $i = 1, \ldots, m$ type of product manufactured
- $k = 1, \ldots, k$ resources provided
- p_{ti} unit production cost per product per period
- q_{ti} fixed production cost per product per period
- h_{ti} cost of storage for each product in each period

- d_{ti} demand for each product in each period
- L_{tk} production capacity of the raw material k in period t
- α_{ik} amount of primary substance use k for product production i
- β_{ik} consumption amount of raw material k for product production i.

5.5.3.2.2 Decision variables

- x_{ti} production volume i in period t
- y_{ti} the binary variable represents the status of the production activity of the product i in period t
- s_{ti} product inventory i in the end of period t.

5.5.3.2.3 Formulation of the model

$$\min \sum_{i=1}^{m} \sum_{t=1}^{n} p_{ti} x_{ti} + q_{ti} y_{ti} + h_{ti} s_{ti}. \tag{6}$$

such that:

$$s_{(t-1)i} + x_{ti} = d_{ti} + s_{ti}, \qquad \forall t,i, \tag{7}$$

$$s_{(0)i} = s_{(n)i} = 0, \qquad \forall i, \tag{8}$$

$$x_{ti} \leq M_{ti} y_{ti}, \qquad \forall t,i, \tag{9}$$

$$\sum_{i=1}^{m} \alpha_{ik} x_{ti} + \sum_{i=1}^{m} \beta_{ik} y_{ti} \leq L_{tk}, \qquad \forall t,k, \tag{10}$$

$$x_{ti} \in \mathbb{R}_{+}, s_{ti} \in \mathbb{R}_{+}, y_{ti} \in \{0,1\}. \tag{11}$$

Equation (6) represents the objective function of minimizing the final cost of multi-product production, considering unit costs, constant production, and storage for each product during a planning period.

Constraint (7) indicates that the total amount of inventory of each product in the warehouse in each period and the amount of demand for it during that period is equal to the total inventory of each product in the warehouse in the previous period and the volume producing that product at that time. According to constraint (8), the model assumes that the inventory of each product in the warehouse before the beginning of the first period and after the end of the planning period is zero. Constraint (9) shows that the volume of production of each product in a period when production activity is stopped must be zero and when the production activity is active, the volume of production of each product can be taken up to an unlimited amount. Limit (10) shows that the total amount of fixed consumption resources and the total consumption of unit production for each product cannot be greater than the total available or available resources for the production of that product. Finally, constraint (11) represents the type of decision variables, so that the volume of production for each product and inventory of each product in the warehouse in each period can be a positive amount of real numbers and how it is active. The manufacturing process for each product in its collection or inactivity is determined by a binary variable. The number 1 indicates the activation of the production process for each product in the period, and the number 0 indicates the inactivation of the process of producing that product in the period.

5.5.3.3 Model 3: The planning model for the distribution stage

5.5.3.3.1 Definitions

- $t = 1, \ldots, n$ planning period
- $i = 1, \ldots, m$ type of product
- p_{ti} cost of distributing each product in each period
- q_{ti} cost of losing demand for each product in each period
- h_{ti} cost of storing for each product in each period
- d_{ti} demand for each product for each period
- L_t distribution capacity per period.

5.5.3.3.2 Decision variables

- x_{ti} volume of product distribution i in period t
- s_{ti} not distributed product inventory i in the end of period t.

5.5.3.3.3 Formulation of the model

$$\min \sum_{i=1}^{m} \sum_{t=1}^{n} p_{ti} x_{ti} + q_{ti} \left(d_{ti} - x_{ti} \right) + h_{ti} s_{ti}. \tag{12}$$

such that:

$$s_{ti} = s_{(t-1)i} + \left(d_{(t-1)i} - x_{(t-1)i} \right), \qquad \forall i, t, \tag{13}$$

$$s_{(0)i} = s_{(n)i} = 0, \qquad \forall i, \tag{14}$$

$$\sum_{i=1}^{m} \left(x_{ti} + \left(d_{(t-1)i} - x_{(t-1)i} \right) \right) \leq L_t, \quad \forall t, \tag{15}$$

$$x_{ti} \in \mathbb{R}_+, s_{ti} \in \mathbb{R}_+. \tag{16}$$

Equation (12) represents the objective function of minimizing the final distribution costs of products, considering distribution costs, loss of order and storage of unpackaged products for each product in each period.

Constraint (13) indicates that the inventory of unpackaged products in the warehouse in each period for each product is equal to the total inventory in the past period with the amount of non-delivered order in the past period. According to constraint (14), the model assumes that the inventory of unpackaged products in warehouse before the beginning of the first period and after the end of the planning period is zero. Constraint (15) indicates that the total amount of product distribution in each period and the amount of unordered orders in the previous period cannot exceed the distribution capacity of the product in that period. Finally, constraint (16) indicates the type of decision variables, so that the volume of distribution of materials and the amount of inventory of unpackaged products in the warehouse will have a value greater than zero in each equilibrium period.

5.6   NUMERICAL RESULTS

According to the description of the operations obtained in the previous section, in this section, with numerical and numerical examples, we analyze the results at all levels of the supply chain turnover structure. For this purpose, using a series of experimental data for each step in the system, we execute each model separately and run it the specified number of times and report the results. Regarding the values of each category of data and the results obtained from them, we report the value of the target function, which is the same amount of cost.

For Model No. 1 in the supply phase, using the four raw materials for purchase and planning for six periods, we use the experimental data group and calculate the results. Figure 5.2 shows the results obtained with respect to the input data for model 1.

Based on the results and calculations shown, the value of the objective function, which shows the total cost of the supply phase, is obtained as follows:

Objective value: 186 085

To analyze the supplier companies and find the optimal result, we will examine the planning of all existing suppliers and compare the value of the target function at each step.

For this set of raw materials and considering 7 parallel suppliers, by entering the information of each supplier, we examine the extent of its objective function and calculate the results. The value of the objective function or the amount of cost to each supplier is as follows.

Supplier Number 1: Objective value: 186 085
Supplier Number 2: Objective value: 171 936
Supplier Number 3: Objective value: 165 892
Supplier Number 4: Objective value: 182 897
Supplier Number 5: Objective value: 163 705
Supplier Number 6: Objective value: 197 570
Supplier Number 7: Objective value: 185 425

Now, for Model No. 2 in the production stage, assuming 4 production and scheduling products for 6 periods, we use the experimental data group and calculate the results. The results of this step are performed according to Model No. 2 presented in the previous chapter. Figure 5.3 shows the results obtained with regard to the input data for Model No. 2.

t	1	2	3	4	5	6
p1	47	50	45	52	39	36
p2	54	47	41	62	67	47
p3	70	53	63	47	45	69
p4	67	54	47	53	52	57
q1	23	17	16	19	18	21
q2	18	16	21	17	19	21
q3	18	23	17	24	22	9
q4	21	18	23	23	19	23
h1	7	7	7	7	7	4
h2	8	4	5	7	4	8
h3	5	6	8	4	7	7
h4	8	5	4	7	6	8
d1	200	130	188	188	161	138
d2	147	185	186	134	132	122
d3	130	133	118	168	136	161
d4	198	136	177	123	157	146
x1	200	130	376	0	161	138
x2	147	185	452	0	0	122
x3	130	251	0	168	297	0
x4	198	136	300	0	157	146
y1	1	1	1	0	1	1
y2	1	1	1	0	0	1
y3	1	1	0	1	1	0
y4	1	1	1	0	1	1
s1	0	0	188	0	0	0
s2	0	0	266	132	0	0
s3	0	118	0	0	161	0
s4	0	0	123	0	0	0

FIGURE 5.2   Output for Model 1.

Based on the results and calculations shown, the value of the objective function, which shows the total cost of the production process, is obtained as follows:

Objective value: 270 763

To analyze the manufacturing companies and find the optimal result, we plan the production of all existing manufacturers for these products, and in each step we compare the value of the target function.

t	1	2	3	4	5	6
p1	44	32	36	33	34	47
p2	50	49	36	33	33	35
p3	47	34	35	44	46	48
p4	36	38	41	47	42	39
q1	18	19	17	19	18	15
q2	18	16	16	15	15	18
q3	16	16	15	19	15	17
q4	16	16	16	18	18	18
h1	7	6	6	8	9	8
h2	5	10	8	7	9	7
h3	6	8	10	10	10	5
h4	8	6	7	7	5	7
d1	278	309	320	338	348	273
d2	305	305	345	304	275	296
d3	269	268	326	302	300	297
d4	330	296	281	330	313	321
I1	6918	6725	6699	6630	6506	6854
I2	6505	6878	6589	6848	6657	6899
x1	502	109	478	126	261	270
x2	277	363	231	325	404	130
x3	254	320	334	314	336	277
x4	259	266	332	308	335	265
y1	1	1	1	1	1	1
y2	1	1	1	1	1	1
y3	1	1	1	1	1	1
y4	1	1	1	1	1	1
s1	167	0	165	0	0	0
s2	0	105	0	0	141	0
s3	0	0	0	0	0	0
s4	0	0	0	0	0	0

-	product1	product2	product3	product4
α_1	5	5	3	6
α_2	3	3	4	6
β_1	94	87	86	87
β_2	92	82	95	100

FIGURE 5.3   Output for Model 2.

For this product category and considering the eight parallel manufacturers, we will examine the amount of target function by entering the information for each producer and compute the results. The value of the objective function or the amount of cost for each producer is as follows.

t	1	2	3	4	5	6
p1	7	10	7	10	6	8
p2	8	8	10	10	10	6
p3	9	6	7	10	7	6
p4	7	6	7	10	10	10
q1	7	4	5	4	7	7
q2	5	4	7	5	5	4
q3	7	6	4	5	7	5
q4	6	4	6	5	5	7
h1	6	6	5	3	5	4
h2	3	4	3	4	3	5
h3	4	3	5	5	4	5
h4	4	6	4	6	4	5
d1	162	152	165	174	219	193
d2	173	217	194	172	205	175
d3	187	204	225	169	232	190
d4	173	191	157	239	219	249
L	986	979	962	1010	1003	971
x1	162	152	165	110	283	193
x2	173	217	194	172	205	175
x3	187	204	225	169	232	190
x4	173	191	157	239	219	249
s1	0	0	0	0	64	0
s2	0	0	0	0	0	0
s3	0	0	0	0	0	0
s4	0	0	0	0	0	0

FIGURE 5.4 Output for Model 3.

Manufacturer Number 1: Objective value: 270 763
Manufacturer Number 2: Objective value: 281 607
Manufacturer Number 3: Objective value: 3 009 914
Manufacturer Number 4: Objective value: 291 555
Manufacturer Number 5: Objective value: 322 762
Manufacturer Number 6: Objective value: 279 521

Manufacturer Number 7: Objective value: 286 820
Manufacturer Number 8: Objective value: 271 738

Now, for Model No. 3 and in the distribution stage, assuming 4 products for distribution and planning for 6 periods, we use empirical data sets and calculate the results. Figure 5.4 shows the results obtained with regard to the input data for model 3.

Based on the results and calculations shown, the value of the objective function, which shows the total cost at the distribution stage, is obtained as follows:

Objective value: 42 315

To analyze distributor companies and find optimal results, we examine the distribution program for all distributors for these products, and in each step we compare the value of the target function.

For this product category and considering 4 parallel distributors, we will examine the amount of the target function by entering the information for each distributor and compute the results. The value of the target function or the amount of cost for each distributor is as follows.

Distributor Number 1: Objective value: 42 315
Distributor Number 2: Objective value: 40 600
Distributor Number 3:Objective value: 45 210
Distributor Number 4:Objective value: 43 720

5.7 DISCUSSION

The main theme of this chapter has been the virtualization of the supply chain management. A brief review on Industry 4.0 areas and their application to supply chain components showed that there is significant value in virtualizing supply chain management. In terms of Industry 4.0, the employment of the IoT has many advantages for services, fast data exchange using Wi-Fi, precise distribution and transportation using GPS and easy access management with web services. Also in cases with larger number of processes and interactions it's complicated to analyze the situation but IoT-based systems can handle the situation. From this point of view, we can recognize that virtualization in supply chain management not only influences cost effectiveness optimization but also rather gives the impression of decreasing risk.

In consideration of this theoretical and conceptual deficit, there might be a risk that Industry 4.0 becomes a so-called management fashion, which is a term referring to management concepts that relatively quickly gain large shares in the public management discourse. However, management fashions usually fail to gain practical relevance in the long run, meaning that they fade under the weight of unfulfilled promises after a period of time. Management fashions often evolve around highly topical issues and result in a substantial number of publications, workshops and conferences. Furthermore, management fashions are frequently used by consulting companies as an instrument to boost demand.

The IoT offers new possibilities in the area of performance. For example, road transport trucks can be automatically controlled according to specifications of users, which will allow them to operate in predefined intervals and with a standard speed, so as to maximize fuel economy. Big data makes it possible to analyze the data at a more advanced level than traditional tools allowed. With this technology, even data which has been collected in various mutually incompatible systems, databases and websites is processed and combined to give a clear picture of the situation when which there is a specific company or person.

Additionally, in comparison with the traditional supply chain, IoT-based system performance is considerable. Actually, an IoT framework represents faster processing that can increase productivity. Furthermore, since conventional data management systems are inefficient in handling large data sets, big data technologies act as a safeguard by helping firms to create real-time intelligence from high volumes of perishable data.

In other words, the result of developing the Internet of Things and big data in supply chain management can be the structure of regional networks, flexibility, risk management and rotation means, increased customer requirements in terms of lead time delivery services, their availability and reliability, services prepared in accordance with the needs of consumers, therefore rapid response to their needs, segmentation of the supply chain focused on demand and specific needs of customers, which can help to reduce the volume of stocks, and thus to optimize costs, safety requirements and potential hazards in the supply chain, risk management in the supply chain, and strategies for sustainable development of enterprises with regard to environmental aspects.

5.8 CONCLUSIONS

Within this chapter, we have shown that the fourth industrial revolution can be viewed as a change in the production logic of increasingly decentralized, self-regulating value creation, capability by concepts and technologies such as CPS, IoT, Big Data, mass calculation, incremental production, and smart factories which are best placed to help companies meet their future production needs. It also provides an overview of the impact that the fourth industrial revolution has on the design and management of the supply chain. In addition, we proposed that companies should be supported in their work in Industry 4.0. This can be achieved through the concepts and frameworks that illustrate the various industry blocks and dimensions of the Industry 4.0 and may thus act as a guide.

For future research, the following could be investigated:

- With respect to the elements and function and their corresponding decisions, mathematical models could be developed as decision aids
- Database clustering and data mining can be embedded through the cloud environment for faster processing
- Integrated multi-objective mathematical models can be formulated for overall decision making in the proposed integrated IoT–based logistics system.

BIBLIOGRAPHY

Baines, T. (2015). *Exploring Service Innovation and the Servitization of the Manufacturing.* Taylor & Francis Group, Firm.

Bauernhansl, T., Ten Hompel, M., & Vogel-Heuser, B. (2014). *Industrie 4.0 in Produktion.* Automatisierung und Logistik: Anwendung, Technologien und Migration: Springer.

Blanchet, M., Rinn, T., Von Thaden, G., et al. (2014). Industry 4.0: The new industrial revolution-How Europe will succeed. Hg. v. Roland Berger Strategy Consultants GmbH. München. Abgerufen am 11.05. 2014, unter www. rolandberger. com/media/pdf/Roland_Berger_TAB_Industry_4_0_2014 0403. pdf

Brettel, M., Friederichsen, N., Keller, M., et al. (2014). How virtualization, decentralization and network building change the manufacturing landscape: An Industry 4.0 Perspective. *International Journal of Mechanical, Industrial Science and Engineering*, 8(1), 37–44.

Chen, H., Chiang, R. H., & Storey, V.C. (2012). Business intelligence and analytics: From big data to big impact. *MIS Quarterly*, 36(4), (December 2012),1165–1188.

Dubey, R., Gunasekaran, A., Childe, S. J., et al. (2016). The impact of big data on world-class sustainable manufacturing. *The International Journal of Advanced Manufacturing Technology*, 84(1–4), 631–645.

Fazlollahtabar, H. (2016). Parallel autonomous guided vehicle assembly line for a semi-continuous manufacturing system. *Assembly Automation*, 36(3), 262–273.

Fazlollahtabar, H. (2018a). Lagrangian relaxation method for optimizing delay of multiple autonomous guided vehicles. *Transportation Letters*, 10(6), 354–360.

Fazlollahtabar, H. (2018b). Scheduling of multiple autonomous guided vehicles for an assembly line using minimum cost network flow. *Journal of Optimization in Industrial Engineering*, 11(1), 185–193.

Fazlollahtabar, H. (2019a). An effective mathematical programming model for production of automatic robot path planning. *The Open Transportation Journal*, 13(1), 11–16.

Fazlollahtabar, H. (2019b). Triple state reliability measurement for a complex autonomous robot system based on extended triangular distribution. *Measurement*, 139, 122–126.

Fazlollahtabar, H. (2020). Comparative simulation study for configuring turning point in multiple robot path planning: Robust data envelopment analysis. *Robotica*, 38(5), 925–939.

Fazlollahtabar, H. (2021). Robotic manufacturing systems using Internet of Things: new era of facing pandemics. *Automation, Robotics & Communications for Industry*, 4.0, 82.

Fazlollahtabar, H. (2022). Internet of Things-based SCADA system for configuring/reconfiguring an autonomous assembly process. *Robotica*, 40(3), 672–689.

Fazlollahtabar, H., & Hassanli, S. (2018). Hybrid cost and time path planning for multiple autonomous guided vehicles. *Applied Intelligence*, 48, 482–498.

Fazlollahtabar, H., & Jalali, S.G. (2013). Adapted Markovian model to control reliability assessment in multiple AGV. *Scientia Iranica*, 20(6), 2224–2237.

Fazlollahtabar, H., & Niaki, S.T.A. (2017a). Binary state reliability computation for a complex system based on extended Bernoulli trials: Multiple autonomous robots. *Quality and Reliability Engineering International*, 33(8), 1709–1718.

Fazlollahtabar, H., & Niaki, S.T.A. (2017b). *Reliability Models of Complex Systems for Robots and Automation.* CRC Press.

Fazlollahtabar, H., & Niaki, S.T.A. (2017c). Integration of fault tree analysis, reliability block diagram and hazard decision tree for industrial robot reliability evaluation. *Industrial Robot: An International Journal*, 44(6), 754–764.

Fazlollahtabar, H., & Niaki, S.T.A. (2018a). Cold standby renewal process integrated with environmental factor effects for reliability evaluation of multiple autonomous robot system. *International Journal of Quality & Reliability Management*, 35(10), 2450–2464.

Fazlollahtabar, H., & Niaki, S.T.A. (2018b). Modified branching process for the reliability analysis of complex systems: Multiple-robot systems. *Communications in Statistics-Theory and Methods*, 47(7), 1641–1652.

Fazlollahtabar, H., Es'haghzadeh, A., Hajmohammadi, H., et al. (2012). A Monte Carlo simulation to estimate TAGV production time in a stochastic flexible automated manufacturing system: a case study. *International Journal of Industrial and Systems Engineering*, 12(3), 243–258.

Fazlollahtabar, H., Mahdavi-Amiri, N., & Muhammadzadeh, A. (2015). A genetic optimization algorithm for nonlinear stochastic programs in an automated manufacturing system. *Journal of Intelligent & Fuzzy Systems*, 28(3), 1461–1475.

Fazlollahtabar, H., Rezaie, B., & Kalantari, H. (2010). Mathematical programming approach to optimize material flow in an AGV-based flexible job shop manufacturing system with performance analysis. *The International Journal of Advanced Manufacturing Technology*, 51(9–12), 1149–1158.

Fazlollahtabar, H., & Saidi-Mehrabad, M. (2015a). Risk assessment for multiple automated guided vehicle manufacturing network. *Robotics and Autonomous Systems*, 74, 175–183.

Fazlollahtabar, H., & Saidi-Mehrabad, M. (2015b). *Autonomous Guided Vehicles: Methods and Models for Optimal Path Planning*. Springer International Publishing, Germany.

Fazlollahtabar, H., & Saidi-Mehrabad, M. (2019). *Cost Engineering and Pricing in Autonomous Manufacturing Systems*. Emerald Publishing Limited.

Fazlollahtabar, H., Saidi-Mehrabad, M., & Balakrishnan, J. (2015a). Mathematical optimization for earliness/tardiness minimization in a multiple automated guided vehicle manufacturing system via integrated heuristic algorithms. *Robotics and Autonomous Systems*, 72, 131–138.

Fazlollahtabar, H., Saidi-Mehrabad, M., & Balakrishnan, J. (2015b). Integrated Markov-neural reliability computation method: A case for multiple automated guided vehicle system. *Reliability Engineering & System Safety*, 135, 34–44.

Fazlollahtabar, H., Saidi-Mehrabad, M., & Masehian, E. (2015). Mathematical model for deadlock resolution in multiple AGV scheduling and routing network: A case study. *Industrial Robot: An International Journal*, 42(3), 252–263.

Fazlollahtabar, H., Saidi-Mehrabad, M., & Masehian, E. (2021). Robotic industrial automation simulation-optimization for resolving conflict and deadlock. *Assembly Automation*, 41(4), 477–485.

Fazlollahtabar, H., & Shafieian, S.H. (2014). An optimal path in an AGV-based manufacturing system with intelligent agents. *Journal for Manufacturing Science and Production*, 14(2), 87–102.

Hermann, M., Pentek, T., & Otto, B. (2016). *Design principles for industrie 4.0 scenarios*. Paper presented at the System Sciences (HICSS), 2016 49th Hawaii International Conference on.

Lachenmaier, J. F., Lasi, H., & Kemper, H.-G. 2015. *Entwicklung und Evaluation eines Informationsversorgungskonzepts für die Prozess-und Produktionsplanung im Kontext von Industrie 4.0. Paper presented at the Wirtschaftsinformatik*.

Perera, C., Ranjan, R., & Wang, L. (2015). Big data privacy in the Internet of Things era. *IT Professional*, 17(3), 32–39.

Qiu, R.G. (2014). *Service Science: The Foundations of Service Engineering and Management*. John Wiley & Sons.

Records, R., & Fisher, Q. (2014). Manufacturers connect the dots with big data and analytics. *Computer Science Corporation*, vol 7, 1–6.

Rennung, F., Luminosu, C.T., & Draghici, A. (2016). Service provision in the framework of Industry 4.0. *Procedia-Social and Behavioral Sciences*, 221, 372–377.

Schmenner, R.W. (2009). Manufacturing, service, and their integration: Some history and theory. *International Journal of Operations & Production Management*, 29(5), 431–443.

Song, M.-L., Fisher, R., Wang, J.-L., et al. (2016). Environmental performance evaluation with big data: Theories and methods. *Annals of Operations Research*, 1–14.

Shirazi, B., Fazlollahtabar, H., & Mahdavi, I. (2010). A six sigma based multi-objective optimization for machine grouping control in flexible cellular manufacturing systems with guide path flexibility. *Advances in Engineering Software* 41(6), 865–873.

Shojaeifar, A., Fazlollahtabar, H., & Mahdavi, I. (2016). Decomposition versus minimal path and cuts methods for reliability evaluation of an advanced robotic production system. *Journal of Automation Mobile Robotics and Intelligent Systems*, 10(3), 52–57.

Tavana, M., Fazlollahtabar, H., & Hassanzade, R. (2014). A Bi-Objective Stochastic programming model for optimizing automated material handling systems with reliability considerations. *International Journal of Production Research* 52(19), 5597–5610.

van der Aalst, W. M., & Arthur, H. (2003). AHM t. Hofstede and M. Weske. "Business Process Management: A Survey," LNCS, 2678: 1–12.

Wahlster, W. (2012). *Industry 4.0: From Smart Factories to Smart Products*. Paper presented at the Forum Business Meets Research BMR.

Wamba, S. F., & Akter, S. (2015). *Big data Analytics for Supply Chain Management: A Literature Review and Research Agenda*. Paper presented at the Workshop on Enterprise and Organizational Modeling and Simulation.

Wang, G., Gunasekaran, A., Ngai, E. W., et al. (2016). Big data analytics in logistics and supply chain management: Certain investigations for research and applications. *International Journal of Production Economics*, 176, 98–110.

6 Sustainable Production in Pandemics

6.1 INTRODUCTION

The United Nations (UN) launched the sustainable development agenda for 2030, which addresses the various on-going challenges related to environmental degradation, climate change, zero hunger, and other negative consequences of the different production processes. Goal-3 of UN 2030 Agenda for sustainable development discusses the "development of healthy lives and promoting wellbeing for all ages". This situation of the pandemic has created new dimensions for the social sustainability of people and manufacturing organizations. Currently, most of the manufacturing and supply chain organizations are struggling to anticipate the negative consequences of COVID-19. Most of the global markets are shrinking, and industrial managers are searching for new materials and process methods to maintain production. Notably, the COVID-19 outbreak significantly improves organizational environmental sustainability, albeit downsizing the consumer economy and raising challenges for the industrial workforce management.

Amidst COVID-19, global supply chain and manufacturing network moved through a very distressing stage. The Emerging Infectious Diseases (EIDs) such as Ebola, influenza, SARS, MERS, and most recently, Coronavirus Disease (COVID-2019) cause enormous disruption in goods production, people's lives, transportation, and stimulate civil unrest. Sustainable development goals are the common agenda for nations to save our planet. The UN SDG outline includes 17 goals with several targets for the operationalization of sustainable development by 2030. The COVID-19 pandemic influenced the global economy, business, socio-economic systems, and human behaviors with different short- and long-term impacts on people, planet, and profit. One of the essential goals in the UN agenda is Goal 12 that seeks sustainable production and consumption. This goal includes different stakeholders and covers the life cycle approach of several sectors including mobility, food, agriculture, housing, and appliances. By June 2021, the pandemic, with more than 177 million cases and 3.8 million deaths, influenced the realization of SDG goals. The main elements in Goal 12, including reducing material consumption and footprint, increasing the recycling rate, decreasing food waste, management of hazardous substances, and increasing the number of policies for promoting sustainable production and consumption. The impacts of the pandemic on these targets should be analyzed by considering

DOI: 10.1201/9781003400585-6

certain scenarios. Several studies recently focused on the impacts of COVID-19 on the supply chain and logistics (Ivanov, 2020; Govindan et al., 2020; Shokrani et al., 2020; Sarkis et al., 2020). A few studies also studied the impacts on some of the SDG goals. However, to the best of our knowledge, addressing the impacts of the pandemic on Goal 12 in a systematic way and via a simulation model was not a prime focus of prior research. The system dynamics approach provides the opportunity to assess a large-scale problem with several influential factors (Giannis et al., 2016). This chapter aims to discuss the application of system dynamics in evaluating the impacts of COVID-19 on Goal 12.

6.2 RELATED WORKS

In this section, a brief review on ensuring a sustainable production and consumption pattern (called Goal 12), the research on sustainable production and consumption, the studies on the impacts of COVID-19 on SDG goals, and the system dynamics approach are provided.

In 2015, the 2030 agenda for sustainable development was agreed upon by all the United Nations members. This agenda with 17 goals provides the shared values for protecting our planet and aids in peace and wealth for all people.[1] One of these goals is ensuring a sustainable production and consumption pattern (Goal 12) with 11 targets. The indicators based on the UNEP[2] source are provided in Figure 6.1. These indicators aid in measuring the achievement of the determined targets to help nations' performance evaluation in the sustainability of production and consumption and highlight the deviations in order to take appropriate action. These indicators include the policies for sustainable procurement, the material footprint, the food wastes, the recycling rate, the hazardous substances, publishing corporate social responsibility reports by companies, and material consumption. The pandemic in 2020 influenced the achievement of these targets. According to the UN report, in 2010, the global material footprint was 73.2 billion metric tonnes that are increased by 17.3% in 2017. From 2017 to 2019, for 79 counties and the EU, at least one policy related to Goal 12 is reported. By 2019, only 20% of electronic wastes are recycled, and these wastes

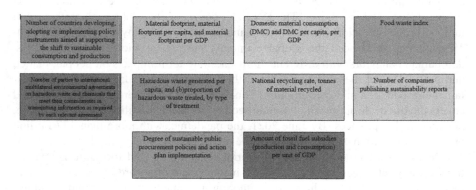

FIGURE 6.1   The indicators for Goal 12 (source of data: UNEP).

have grown by 38%. Despite the target of reducing the fuel subsidies, they increased by 34% from 2015 to 2018. The food wastes through the supply chain stream were reported at around 13.8% in 2016 (Keivanpour, 2015).

Pradhan et al. (2017) performed a systematic review on the interactions between SDG goals. They considered the positive correlation as synergy and the negative as a trade-off. Based on the indicators and the data of 227 countries, they concluded that Goal 1 (no poverty) has the most synergetic relations with the other SDG goals, and Goal 12 is most linked to the trade-offs. Hence, for trade-offs, indicators and targets should be well discussed and negotiated for achieving better outcomes of these strategies (Pradhan et al., 2017). Sala and Castellani (2019) used a life cycle assessment approach for evaluating the environmental impacts of food, mobility, housing, household goods, and appliances. They concluded that food is the most important area for consideration. The use phase plays an essential role in mobility, appliances, and consumption housing. For food and household goods, the upstream of the supply chain plays an essential role in environmental impacts. Gasper et al. (2019) discussed Goal 12, related targets, indicators, and policies. They mentioned that several challenges existed related to the indicators and monitored the target across the countries. They suggested developing National targets for improving the outcomes. Chan et al. (2018) reviewed Goal 12 and they highlighted several gaps including focusing on the material footprint by reducing the overall consumption, improving the knowledge of decision-makers, policymakers, consumers, and businesses on sustainable production and consumption.

A few studies recently considered the impacts of the pandemic on Goal 12. Elavarasan et al. (2021) studied the impacts of the pandemic on Goal 7 (energy sustainability). They used Analytic Hierarchy Process (AHP) and Strength, Weakness, Opportunity, Threat (SWOT) analysis to address the post-pandemic scenarios. They also discussed the impacts of COVID-19 on all 17 SDG goals and the challenges and solutions of the energy sector. For Goal 12, they mentioned the availability of resources in healthcare systems and hospitals and the medical wastes being generated during the pandemic as an important challenge. They suggested the optimization of consumption and production, promoting the 3R approach (reducing, reusing, recycling) and using the sharing platforms as the solutions for dealing with challenges. Nundy et al. (2021) studied the impacts of the pandemic on energy, transport, and socioeconomic sectors. They mentioned decreasing the energy demand in the transport sector and revenue loss while increasing energy consumption in the residential sector. The pandemic influenced SDG target achievement by 2030, and a collaborative mitigation plan is required. Health also considered the impacts of the pandemic on SDG goals. The author discussed that achieving goals 7–9 and 11–15 with the long-term impacts of COVID-19 is challenging. The economic recession affects the policies and the implementation of SDG goals. Several studies addressed the impacts of the pandemic on the supply chain and logistics. Kumar et al. (2020) discussed the impacts of the pandemic on sustainable production and consumption. They recommended some policies in this area. For production, the policies should be revised to consider the impacts of COVID-19 with appropriate incentives. Industry 4.0 and digital manufacturing should be developed. The coordination between different stakeholders should be performed. The real-time monitoring and control of production could reduce the impacts of disruption. For consumption, e-commerce is growing considerably during the lockdowns. The role of social media is essential in evaluating consumers' behavior. The uncertainties of consumer demand should be managed with the resilience of the supply chain and certain strategies for risk management.

System dynamics is a powerful approach for assessing the complexity and the interaction of the dynamics models. This approach is used in the context of sustainable development considering the complexity, the role of different stakeholders, and the variation in the influential variables. Hjorth et al. (2006) used system dynamics in the context of sustainable development. They mentioned complexity and the self-organizing system as the characteristics for using this approach. Giannis et al. (2016) used system dynamics for evaluating different recycling scenarios. They applied this approach due to several influential factors on waste management including population, rapid economic growth, and consumer patterns.

The literature highlights the following points:

- Goal 12 of SDG plays an essential role in sustainability and more research is required considering the trade-offs, the complexity, and the challenges of the current indicators.
- The pandemic affects the realization of SDG goals by 2030. A few studies discussed the impacts on the goals. However, focusing on Goal 12 and the related indicators has not received much attention in the literature.
- System dynamics is an effective approach for addressing the complexity and self-organizing nature of sustainable development problems. This approach could aid in analyzing the impacts of COVID-19 on the key elements of Goal 12.

6.3 PROPOSED APPROACH

The pandemic situation gives rise to a demand for rare production items such as ventilators, gloves, face shields, masks, and sanitizers at a high rate. During this pandemic era, some of the manufacturing giants such as General Motors and Ford Motors turn their production system to support the need of society in terms of manufacturing ventilators. Therefore, a flexible manufacturing system is required to fulfil the requirement for such necessary items. National government institutions, manufacturing organizations, and health institutions should be prepared in advance to tackle the pandemic situation to control the production of essential and nonessential items during a pandemic. This means that they should have sufficient buffer plans to address the availability of life saving stock such as ventilators, vaccines, sanitizers, masks, and face shields.

The post COVID era opens an opportunity window for a sustainable business transition, and need to make supply and production system more resilient. The COVID-19 situation creates a space for developing a flexible and resilient manufacturing system to maintaining the economic and social sustainability of the production process. Tan et al. (2019) discuss the various decision support systems for developing a resilient production system. Ivanov (2020) proposed a prediction model for measuring the impact of a pandemic on the supply chain network and manufacturing resilience. The firm's supply chain network resilience and manufacturing resilience is required to tackle the epidemic or or other such disruptive events. Due to such disruptive events, material shortage and delivery delays are seen in the downstream supply chain, causing a ripple effect and resulting in reduced performance in terms of service level, revenue, and process productivity.

6.3.1 Nature and Impact of Pandemic

The spread of COVID-19 has already disrupted supply chains globally in several ways, including:

- Weakening demand for several types of products (automobile products, public transport, and textile products)
- Skyrocketing demand for select companies and their products (thermal scanners, ventilators, face masks, sanitizers, PPE, and essential food items)
- Failure of supplies and uncertainty in raw material supply
- Impacting the ability to ship and receive products on time due to shortages and logistics bottlenecks
- Ensuring workforce capacity to assemble and ship products.

To deal with above-mentioned issues, a smooth production plan is required for the pandemic period. This involves the need for manufacturing firms to be more resilient and flexible to produce the essential items to meet customer demand. Artificial Intelligence (AI)-based production technologies are recommended to produce granular COVID-19 maps that enable individuals to decide what places to avoid, companies to manage risk, and governments to deploy resources. AI-based manufacturing is an excellent option to promote digital production during pandemics.

Collaborating with government agencies is required to develop drugs to reduce the severity of COVID-19 illness and treat infections, possibly leading to a vaccine. The application of advanced technologies such as AI, 3D printing, data analytics, Robots, cyber-physical systems can help in developing a decentralizing production system. Digital manufacturing would help to maintain social distancing during the production process, and consequently control the worker's movement. For instance, Kiva robots are used for warehousing to keep the items in their places and help in loading and unloading. Many hospitals are using robots for delivering drugs and monitoring COVID-19 patients. Similarly, the food and essential items delivered in cities can be regulated using technology such as drones, and consequently, prevent direct contact in the delivery of the products.

In this chapter, first, the essential indicators of Goal 12 related to production and consumption are addressed. For this study, we focus on domestic material consumption, the material footprint, food wastes, and the recycling rate. The software Vensim PLE was used for designing a system dynamic model. Figure 6.2 shows

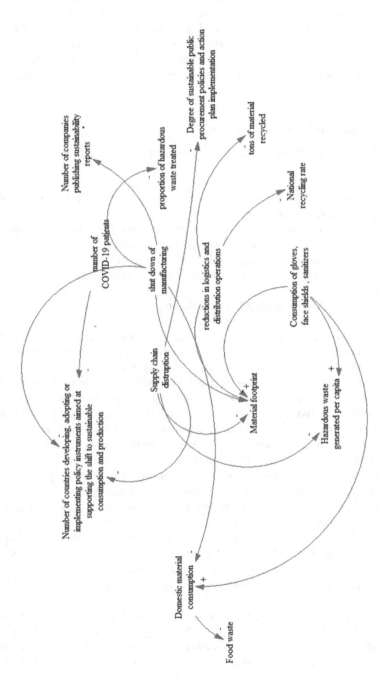

FIGURE 6.2 Causal loop diagram model for the impacts of the pandemic on indicators.

the causal loop diagram. The increasing number of COVID-19 patients caused several shutdowns of manufacturing units and led to supply chain disruption, reduction in logistics operations, and transport. The resultant disruption decreased the allocated resources for the enterprises on sustainable procurement and publishing of the Corporate Social Responsibilities (CSR) reports. The priorities for countries were shifted as the results of the loads on healthcare systems and hospitals. Hence, it decreased the number of developed and adopted policies in sustainable production and consumption. The disruption in the logistics and supply chain also caused a decrease in the material footprint and hazardous wastes generated from factories. The lockdowns of restaurants and the cities contributed to decreasing food wastes. However, the impacts on the wastes through the supply chain require detailed data. The huge consumption of medical protection items such as glasses, face shields, surgical masks, and gloves generated a considerable amount of solid waste. The recycling, landfilling, and proper treatment of these wastes should be addressed in the related policies. The main dynamic equations in this study are presented in this section. Equations (1)–(5) show the system equations based on UNEP metadata of the Goal 12 indicators. Three elements are essential in domestic material consumption: direct import material, domestic extraction, and direct exports. For material footprint, the raw material of import, domestic extraction, and raw material of exports should be considered. Food wastes are the result of two indexes: food waste proportion to total wastes and the food wastes during different supply chain tiers. The other indicator is recycling rate, which is a function of material recycled, material exported and imported for recycling, and total wastes. Figure 6.3 shows the system dynamic model including five subsystems.

$$\text{DMC} = MIm + MExt - Mexp \tag{1}$$

$$MF = RMIm + MExt - RMex \tag{2}$$

$$Foodwastes = \{Fw/T\_W, \Sigma ni = 1 Fwtieri\} \tag{3}$$

$$Rrate = (Mrecy + Mexp\text{-}recy - MImp\text{-}recy)/T\_W \times 100 \tag{4}$$

$$TW = Wman + WEcon + We_{,,,}\text{-}supply + Wmun \tag{5}$$

The first sub-system is domestic material consumption. The second is the material footprint. The national recycling rate, total wastes, and the food wastes index are the third, fourth, and fifth sub-systems in the proposed model, respectively. A scenario generation process based on the COVID-19 data could be designed to address the different states. For example, Ivanov (2020) used three scenarios for addressing the impacts of COVID-19 on the supply chain disruption (S1: outbreak in China, S2: outbreak in China, US and Europe, and S3: epidemic crisis). The severity and the duration could be added for building different scenarios. Table 6.1 shows the examples of the scenarios based on three factors of the region: outbreak, severity, and duration.

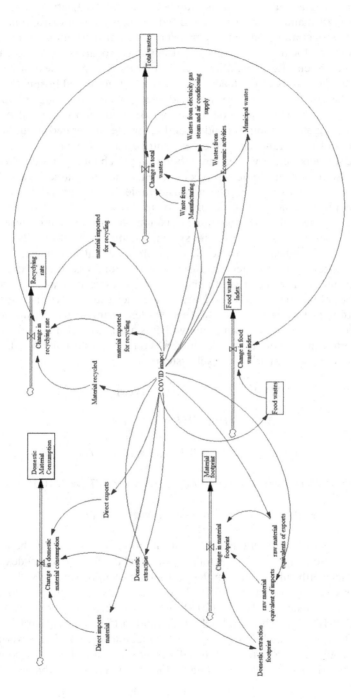

FIGURE 6.3 The system dynamic model with five sub-systems.

TABLE 6.1
The examples of the generated scenarios

Scenarios	Regions' Outbreak	Severity	Duration
1	Asia	High	3 months
2	Asia + US and Europe	High	6 months
3	Asia	Moderate	3 months
4	Asia + US and Europe	Moderate	3 months

Abbreviations:

DMC	Domestic material consumption
MIm	Direct imports material
MExt	Domestic extraction
Mexp	Direct exports
MF	Material footprint
RMIm	Raw material equivalent of imports
RMex	Raw material equivalents of exports
Foodwastes	Food wastes in the waste stream and in the supply chain
Fw	Food wastes
T_W	Total wastes
Fwtieri	Food wastes generated in tier i of the supply chain
Rrate	Recycling rate
Mrecy	Material recycled
Mexp-recy	Material exported for recycling
MImp-recy	Material imported for recycling
Wman	Waste from Manufacturing
WEcon	Waste from other economic activities
We,g,s,a-supply	Waste from electricity, gas, steam, and air conditioning supply
Wmun	Municipal Wastes

6.4 MANAGERIAL IMPLICATIONS

One of the main challenges in performance evaluation based on SDG goals is the availability of data. Tracking and tracing this data is challenging due to the complexity of the supply chain and the interaction among the sectors. For example, measuring the wastes in different food supply chain tiers is challenging. Hence, for facilitating accessing data, a country's data is proposed for testing and validating the model. The uncertainties in data are related to the randomness of a factor and the lack of knowledge (Pishvaee et al., 2010). In this context, using a fuzzy number is promising for the design of the experiment. Hence, using the data of Canada and applying fuzzy numbers for some of the variables is the authors' agenda for the completion of this work in progress research.

6.4.1 RECOMMENDATIONS IN THE PRODUCTION DOMAIN FOR POLICY MAKERS AND PRACTITIONERS

- The global and national production policies should be revised. The government needs to support the production system by providing adequate incentives in future policies.
- Current production facilities should shift to digital manufacturing (or Industry 4.0-based manufacturing), and promote the digital technologies such as AI, 3D printing, robots, cyber-physical systems, digital manufacturing, blockchain, and the like, for production of goods.
- A strong coordination mechanism is required among and between stakeholders such as government, manufacturers, medical institutions, NGOs, and possibly military agencies to better control the infection rate of such a pandemic.
- The present pandemic situation would boost the application of digital manufacturing in the healthcare and FMCG sector.
- Pandemic control can be handled with the adoption of a robust information technology management system to share the real-time production and consumption patterns.

The evaluation phase of a pandemic is the most critical and needs proper mitigation strategies. The evaluation of any epidemic can be viewed from the perspective of future manufacturing strategies adoption, contract policies, and network design to support the industry viability. Due to COVID-19, most of the global or local manufacturing industries (automobile, transport, pharmaceutical, food, and the like.) need to revive their production capacity and raw material sourcing. The revamping of the industry also needs further support from national and regional governments. For instance, the Indian government is aiming to boost their self-reliance in the manufacturing sector and extend their support to the Micro, Small, and Medium Enterprises (MSMEs) through various financial and nonfinancial incentives. COVID-19 is significantly slowing down the global trade (including transport and manufacturing) of the various countries because most of the trade routes are disruptive or diverted due to the pandemic. In view of such impacts, this is the right time for industries to adopt and implement Industry 4.0 and digital technologies in manufacturing. The deployment of robots in the medical system can reduce the risk of COVID-19 spread and ensures better monitoring of patients.

The service supply chain is highly impacted due to the spread of COVID-19. Most of the service industries, such as logistics, hospitality, restaurants, and tourism, observed a reduction in demand. Service organisations should use digital technologies for handling customer services. The the implementation of digital technologies in the service industry helps to reduce the chances of the contagion of the COVID-19.

The manufacturing plants should shift their manufacturing capabilities to digital manufacturing to reduce the numbers in their workforces and consequently reduce the chances of worsening the pandemic situation. The transportation industry faces a severe shortage of drivers and vehicle connectivity. Therefore an optimised supply chain network is required to serve more manufacturing plants. This chapter recommends various strategies to improve the resilience and sustainability of the production and logistics process.

The pandemic situation reduces global carbon emissions and improves environmental sustainability. However, at the same time, businesses and people all across the globe are struggling in terms of job downsizing, workers' safety and mental health issues, and financial burdens due to production losses and closure of sites. In view of these considerations, this chapter aims to address critical aspects of sustainable production and consumption of products and services in the context of COVID-19. This chapter also advises future researchers to identify the critical success factors, inhibitors, and drivers to handle a pandemic situation and proposes policy frameworks for improving the resilience of production and operations processes.

6.5 CONCLUSIONS

The operationalization of sustainable development in supply chains and logistics is the key issue in the realization of the United Nations (UN) goals. Goal 12 addresses responsible production and consumption. It includes the sustainable use of natural resources, waste management, applying sustainability through 3R (Reduce, Reuse, and Recycle), and management of the product's life cycle. According to the recent report of the UN, we still face challenges in achieving sustainable production and consumption. The consumption of natural resources is not sustainable. The recycling rate of the consumed products, particularly the electronic products, is far from the target rate. The wastes in the food supply chain during harvesting, transportation, processing, and storage are considerable. Reduction in oil prices and fossil fuel subsidies lead to more emissions and climate change. The technology revolution could play an essential role as leverage in achieving sustainable development goals. The UN's sustainable development goals are interconnected. Moreover, sustainable production and consumption and Industry 4.0 play an essential role in future city planning. Smart sustainable cities are an emerging research theme that addresses the main concerns in planning, transportation, energy consumption, and the risks related to climate change in the new digitalization era. The pandemic in 2020 changed global production and consumption patterns and affected the strategies and policies in achieving the SDG goals. The system dynamics approach provides a simulation environment for analyzing the different scenarios. The multiple feedback relationship between material consumption, material footprint, recycling rate, total wastes, and the food wastes index sub-systems are designed to permit analyzing different scenarios and the impacts on Goal 12. This visualization tool aids decision-makers and policy planners in developing effective strategies and new indicators in the context of SDG. In this study, we focused on five essential subsystems on the basis of Goal 12 indicators. However, considering the subsystem related to the number of countries adopting and implementing policies related to sustainable production and consumption, hazardous waste management, publishing CSR reports, and fossil fuel subsidies are also proposed as future research.

NOTES

1 The Sustainable Development Goals, United Nations. Sustainable Development Goals. 2018. Available online: www.un.org/development/desa/publications/the-sustainable-deve lopment-goals-report-2018.html
2 UNEP. Available online: www.unep.org/explore-topics/sustainable-development-goals/ why-do-sustainable-development-goals-matter/goal-12

BIBLIOGRAPHY

Chan, S., Weitz, N., Persson, Å., et al. (2018). *SDG 12: Responsible Consumption and Production; A Review of Research Needs*. Technical Annex to the Formas Report Forskning för Agenda 2030; Stockholm Environment Institute: Stockholm, Sweden.

Elavarasan, R.M., Pugazhendhi, R., Jamal, T., et al. (2021). Envisioning the UN Sustainable Development Goals (SDGs) through the lens of energy sustainability (SDG 7) in the post-COVID-19 world. *Applied Energy*, 292, 116665.

Fazlollahtabar, H. (2016). Parallel autonomous guided vehicle assembly line for a semi-continuous manufacturing system. *Assembly Automation*, 36(3), 262–273.

Fazlollahtabar, H. (2018a). Lagrangian relaxation method for optimizing delay of multiple autonomous guided vehicles. *Transportation Letters*, 10(6), 354–360.

Fazlollahtabar, H. (2018b). Scheduling of multiple autonomous guided vehicles for an assembly line using minimum cost network flow. *Journal of Optimization in Industrial Engineering*, 11(1), 185–193.

Fazlollahtabar, H. (2019a). An effective mathematical programming model for production of automatic robot path planning. *The Open Transportation Journal*, 13(1), 11–16.

Fazlollahtabar, H. (2019b). Triple state reliability measurement for a complex autonomous robot system based on extended triangular distribution. *Measurement*, 139, 122–126.

Fazlollahtabar, H. (2020). Comparative simulation study for configuring turning point in multiple robot path planning: Robust data envelopment analysis. *Robotica*, 38(5), 925–939.

Fazlollahtabar, H. (2021). Robotic manufacturing systems using Internet of Things: new era of facing pandemics. *Automation, Robotics & Communications for Industry*, 4.0, 82.

Fazlollahtabar, H. (2022). Internet of Things-based SCADA system for configuring/reconfiguring an autonomous assembly process. *Robotica*, 40(3), 672–689.

Fazlollahtabar, H., & Hassanli, S. (2018). Hybrid cost and time path planning for multiple autonomous guided vehicles. *Applied Intelligence*, 48, 482–498.

Fazlollahtabar, H., & Jalali, S.G. (2013). Adapted Markovian model to control reliability assessment in multiple AGV. *Scientia Iranica*, 20(6), 2224–2237.

Fazlollahtabar, H., Mahdavi-Amiri, N., & Muhammadzadeh, A. (2015). A genetic optimization algorithm for nonlinear stochastic programs in an automated manufacturing system. *Journal of Intelligent & Fuzzy Systems*, 28(3), 1461–1475.

Fazlollahtabar, H., & Niaki, S.T.A. (2017a). Binary state reliability computation for a complex system based on extended Bernoulli trials: Multiple autonomous robots. *Quality and Reliability Engineering International*, 33(8), 1709–1718.

Fazlollahtabar, H., & Niaki, S.T.A. (2017)b. Integration of fault tree analysis, reliability block diagram and hazard decision tree for industrial robot reliability evaluation. *Industrial Robot: An International Journal*, 44(6), 754–764.

Fazlollahtabar, H., & Niaki, S.T.A. (2017c). *Reliability Models of Complex Systems for Robots and Automation*. CRC Press.

Fazlollahtabar, H., & Niaki, S.T.A. (2018a). Cold standby renewal process integrated with environmental factor effects for reliability evaluation of multiple autonomous robot system. *International Journal of Quality & Reliability Management*, 35(10), 2450–2464.

Fazlollahtabar, H., & Niaki, S.T.A. (2018b). Modified branching process for the reliability analysis of complex systems: Multiple-robot systems. *Communications in Statistics-Theory and Methods*, 47(7), 1641–1652.

Fazlollahtabar, H., & Saidi-Mehrabad, M. (2015a). *Autonomous Guided Vehicles: Methods and Models for Optimal Path Planning*. Springer International Publishing, Germany.

Fazlollahtabar, H., & Saidi-Mehrabad, M. (2015b). Risk assessment for multiple automated guided vehicle manufacturing network. *Robotics and Autonomous Systems*, 74, 175–183.

Fazlollahtabar, H., & Saidi-Mehrabad, M. (2019). *Cost Engineering and Pricing in Autonomous Manufacturing Systems*. Emerald Publishing Limited.

Fazlollahtabar, H., Saidi-Mehrabad, M., & Balakrishnan, J. (2015a). Mathematical optimization for earliness/tardiness minimization in a multiple automated guided vehicle manufacturing system via integrated heuristic algorithms. *Robotics and Autonomous Systems*, 72, 131–138.

Fazlollahtabar, H., Saidi-Mehrabad, M., & Balakrishnan, J. (2015b). Integrated Markov-neural reliability computation method: A case for multiple automated guided vehicle system. *Reliability Engineering & System Safety*, 135, 34–44.

Fazlollahtabar, H., Saidi-Mehrabad, M., & Masehian, E. (2015a). Mathematical model for deadlock resolution in multiple AGV scheduling and routing network: A case study. *Industrial Robot: An International Journal*, 42(3), 252–263.

Fazlollahtabar, H., Saidi-Mehrabad, M., & Masehian, E. (2015b). Mathematical model for deadlock resolution in multiple AGV scheduling and routing network: a case study. *Industrial Robot: An International Journal*, 42(3), 252–263.

Fazlollahtabar, H., Saidi-Mehrabad, M., & Masehian, E. (2015c). Mathematical model for deadlock resolution in multiple AGV scheduling and routing network: A case study. *Industrial Robot: An International Journal*, 42(3), 252–263.

Fazlollahtabar, H., Saidi-Mehrabad, M., & Masehian, E. (2021). Robotic industrial automation simulation-optimization for resolving conflict and deadlock. *Assembly Automation*, 41(4), 477–485.

Fazlollahtabar, H., & Shafieian, S.H. (2014). An optimal path in an AGV-based manufacturing system with intelligent agents. *Journal for Manufacturing Science and Production*, 14(2), 87–102.

Gasper, D., Shah, A., & Tankha, S. (2019). The framing of sustainable consumption and production in SDG 12. *Global Policy*, 10, 83–95.

Giannis, A., Chen, M., Yin, K., et al. (2016). Application of system dynamics modeling for evaluation of different recycling scenarios in Singapore. *Journal of Material Cycles Waste Management*, 19, 1177–1185.

Govindan, K., Mina, H., & Alavi, B. (2020). A decision support system for demand management in healthcare supply chains considering the epidemic outbreaks: A case study of coronavirus disease 2019 (COVID-19). *Transportation Research Part E: Logistics and Transportation Review*, 138, 101967.

Hjorth, P., & Bagheri, A. (2006). Navigating towards sustainable development: A system dynamics approach. *Futures*, 38, 74–92.

Ivanov, D. (2020). Predicting the impacts of epidemic outbreaks on global supply chains: A simulation-based analysis on the coronavirus outbreak (COVID-19/SARS-CoV-2) case. *Transportation Research Part E: Logistics and Transportation Review*, 136, 101922.

Keivanpour, S. (2015). *An Integrated Approach to Value Chain Analysis of End of Life Aircraft Treatment*. Ph.D. Thesis, Université Laval, Québec, QC, Canada,.

Keivanpour, S. (2022). The impacts of the pandemic on sustainable production and consumption: Toward a system dynamics approach. *Environmental Sciences Proceedings*, 15, 35.

Kumar, A., Luthra, S., Kumar Mangla, S. et al, (2020). COVID-19 impact on sustainable production and operations management, *Sustainable Operations and Computers*, 1, 1–7.

Kumar, A., Luthra, S., Mangla, S.K., et al. (2020). COVID-19 impact on sustainable production and operations management. *Sustainable Operations and Computers*, 1, 1–7.

Nundy, S., Ghosh, A., Mesloub, A., et al. (2021). Impact of COVID-19 pandemic on socio-economic, energy-environment and transport sector globally and sustainable development goal (SDG). *Journal of Cleaner Production*, 312, 127705.

Pishvaee, M.S., & Torabi, S.A. (2010). A possibilistic programming approach for closed-loop supply chain network design under uncertainty. *Fuzzy Sets and Systems*, 161, 2668–2683.

Pradhan, P., Costa, L., Rybski, D., et al. (2017). A systematic study of sustainable development goal (SDG) interactions. *Earths Future*, 5, 1169–1179.

Sala, S., & Castellani, V. (2019). The consumer footprint: Monitoring sustainable development goal 12 with process-based life cycle assessment. *Journal of Cleaner Production*, 240, 118050.

Sarkis, J., Cohen, M.J., Dewick, P., et al. (2020). A brave new world: Lessons from the COVID-19 pandemic for transitioning to sustainable supply and production. *Resources, Conservation and Recycling*, 159, 104894.

Shojaeifar, A., Fazlollahtabar, H., & Mahdavi, I. (2016). Decomposition versus minimal path and cuts methods for reliability evaluation of an advanced robotic production system. *Journal of Automation Mobile Robotics and Intelligent Systems*, 10(3), 52–57.

Shokrani, A., Loukaides, E.G., Elias, E., et al. (2020). Exploration of alternative supply chains and distributed manufacturing in response to COVID-19; A case study of medical face shields. *Master of Design*, 192, 108749.

Tan, R.R., Promentilla, M.A.B., & Tseng, M.L. (2019). Special issue: Decision support for sustainable and resilient systems. *Sustainble Production and Consumption*. 10.1016/j.spc.2019.08.006.

7 Industrial IoT in Pandemics

7.1 INTRODUCTION

The Internet of Things (IoT) has been extracted as a useful tool in Industry 4.0. The IoT in manufacturing systems enables effective process managers to monitor and supervise production using dispatching rules. One of the most efficient production systems is Robotic One. Robots are able to process manufacturing tasks faster and with higher quality without working time constraints. The challenge is in the robot control system leading to real time decisions. To handle large amounts of data on manufacturing floor and in production using robot real time control, the IoT is employed to deliver a mechanism through internet-oriented technologies. The robotic manufacturing system is highly flexible to satisfy customized production according to customers' demands. Automation is cost-effective due to higher throughput and productivity. Thus, in this chapter we aim to propose a comprehensive model to control IoT robotic manufacturing in pandemics to prevent lost sales and economic loss and to keep the sustainable production (Alcácer and Cruz-Machado, 2019).

The interrelationships of modern manufacturing systems with the promotion of Information Technology (IT) related innovations leads to prompt data-driven decision making in all levels of a hierarchical production system based on real-time data exchange systems. In the literature, the development of manufacturing systems is widely correlated with the advancement of IT; for instance, Computer Numerical Control (CNC) collaborating with industrial robots evolved Flexible Manufacturing Systems (FMSs) and so forth. The adoption of IT software development services convinced enterprises to raise their conventional manufacturing systems to a higher level with respect to the IT infrastructure provided (Ju and Son, 2020; Mahbub, 2020).

The concept of smart manufacturing systems prompts within the Industry 4.0 paradigm the employment of cyber-physical systems and the Internet of Things to coordinate the manufacturing functions and transfer the required data for appropriate and effective decentralized decision-making. The communication, cooperation and coordination among entities themselves and human operators are the major key performance criteria to evaluate the success factors of the system (Valecce et al., 2019). Robotics and the IoT communicate within smart manufacturing to configure the Internet of Robotic Things (IoRT) paradigm. In the IoRT, intelligent robots can sense occurrences around a certain domain of distance and transfer information through the internet to a control and decision processing unit and distribute the data among other entities and then decide about appropriate action.

The vision of the manufacturing Internet of Things is based on the notion of Industry 4.0 that denotes technologies and concepts related to Cyber-Physical Systems (CPSs) and the Internet of Things (IoT). In modern manufacturing, CPSs monitor physical processes, create a virtual copy of the physical world and make decentralized decisions. In addition, manufacturing systems reveal increasing characteristics of discretization, intelligence and autonomy. With the complexity of manufacturing systems, it becomes very difficult to realize these characteristics by traditional technologies. Currently, the research around the IoT in manufacturing systems is mainly aimed at cloud manufacturing systems (Li et al., 2010; Tao et al., 2014). This type of manufacturing model transforms traditional product oriented manufacturing to service oriented manufacturing. Research around CPS in manufacturing systems is primarily focused on the communication between physical entities (Shi et al., 2011). It pays more attention to artificial intelligence, adaptability, self-organization, self-regulation and other aspects of autonomic computing functions embedded in physical systems.

Digitalization changes all areas of life and existing business models. This increases pressure on the industry, but opens up new business opportunities at the same time. Industrial enterprises need to reduce time to market significantly, massively improve flexibility, and increase quality while reducing energy and resource consumption while maintaining an acceptable level of security (Mourtzis et al., 2016).

Major challenges of manufacturing related to the Internet of Things are:

- Reducing time-to-market: Due to faster-changing consumer demands, manufacturers have to launch products more quickly, despite rising product complexity. Traditionally, the big competitor has beaten the small one, but now the faster one is beating the slower.
- Enhancing flexibility: Consumers want individualized products, but at the prices they'd pay for mass-produced goods. As a consequence, production has to be more flexible than ever before.
- Increasing quality: Consumers reward high quality by recommending products on the Internet, and they punish poor quality the same way. To ensure high product quality and to fulfill legal requirements, companies have to install closed-loop quality processes, and products have to be traceable.
- Increasing efficiency: Today, it is not only the product that needs to be sustainable and environmentally friendly, but also energy-efficiency in manufacturing and production becomes a competitive advantage, too.
- Increasing security: Another general requirement is security. Digitalization also leads to increasing vulnerability of production plants to cyber-attacks, and this increases the need for appropriate security measures.

In popular parlance, the IoT refers to "things" such as Nest thermostats, glucose-monitoring devices, cars with collision-detecting sensors, even pets with embedded microchips, that relay information about themselves via the internet to someone, or more accurately, to some computer, that uses the information in an intelligent way. The IoT is not limited to consumer products. The aviation industry, for instance, is undergoing an IoT-based revolution by harnessing information generated by engine sensors. Engine manufactures now have access to enormous amounts of data about engines in flight, which they are using to find ways to reduce fuel consumption and become aware of anomalies while a flight is in progress rather than after the fact (Rymaszewska et al., 2017). This is changing entire business models, with Rolls Royce and GE now contracting for hours of operation, not sales of engines.

7.2   RELATED WORKS

Over the IoT, cyber-physical systems communicate and cooperate with each other and humans in real time. In parallel, the advancement of IoT and CPS brings some challenges to manufacturing systems. Research on manufacturing systems is focused on production scheduling (Burnwal and Deb, 2013; Gen and Lin, 2014; Erol et al., 2012), production control (Ounnar and Pujo, 2012; Njike et al., 2012), production management (Ngai et al., 2012), and other aspects of the manufacturing industry.

The IoT is democratizing equipment automation and access to equipment data. It is enhancing flexibility in measurement and actuation possibilities across the shop-floor and freeing manufacturers from the time and costs associated with changes to complex interfaces that, up until now, were only available on sophisticated machinery. Advances in miniaturized electronics have meant that computing and communication power can be added to everything on the shop floor; enabling wide-spread, easy to achieve and economical automation of individual products, materials and machines (Thramboulidis and Christoulakis, 2016). Even legacy equipment that does not support communication interfaces can be integrated by adding sensors placed around the plant and on machines.

In the past year or so, numerous studies have discussed the key potentials of IoT and other digital technologies in the fight against COVID-19 or future pandemics. Elansary et al. (2021) presented a survey of IoT-based solutions used to fight COVID-19. Brem et al. (2021) analyzed the effect of this pandemic on various technologies and discussed their social impacts. A detailed review of digital health solutions used in countries with high COVID-19 cases is presented in (Kalhori et al., 2021). A consensus of Chinese experts on IoT-aided diagnosis of COVID-19 and its treatment is presented in (Bai et al., 2020). Impacts of IoT implementation in healthcare in terms of cost, time and efficiency are enlisted in (Singh et al., 2020). Applications of the IoT, big data, Artificial Intelligence (AI) and blockchain in mitigating the impact of COVID-19 are explored in (Ting et al., 2020). Rahman et al. (2020) proposed some IoT applications that can be useful in combatting COVID-19. Javaid et al. (2020) discuss how different Industry 4.0 technologies (for example, AI, the IoT, virtual reality, and the like.) can help reduce the spread of disease. Applications of AI for COVID-19 have been proposed in (Vaishya et al., 2020). A comprehensive review of the COVID-19 pandemic and the role of IoT, drones, AI, blockchain, and 5G in managing its impact is explored in (Chamola et al., 2020). Mokbel et al. (2020) argued that contact tracing should be the responsibility of facilities and propose a contact tracing architecture which is fully automated and does not depend on user cooperation. Nasajpour et al. (2020) discussed several IoT healthcare applications during three main phases: early diagnosis, quarantine time, and after recovery. A survey by Nayak et al. (2021) discussed the use of Machine Learning (ML), AI and other intelligent approaches for the prognosis of COVID-19.

Dong and Yao (2020) have compiled potential IoT-based solutions to combat COVID-19. They present a detailed study on the capabilities of existing IoT systems at different layers such as the perception layer, the network layer, the fog layer and the cloud layer. Moreover, they also discuss applications of IoT in diagnosing symptoms of COVID-19. A four-layered architecture based on IoT and blockchain technologies has been proposed in (Alam, 2020) to help the fight against COVID-19. The blockchain-based method was proposed to ensure privacy and security of physiological information shared among IoT nodes. It also enlists various applications that have been developed for detecting and tracing potential COVID-19 patients. The role of the IoT in existing digital healthcare infrastructure has been discussed in (Kelly et al., 2020). It also debates the implications of data generated through IoT enabled healthcare infrastructure on the decisions made by policy makers. Moreover, existing enablers and barriers in adopting IoT-based healthcare have also been enlisted.

A detailed survey of the contributions of the IoT in healthcare in response to COVID-19 is provided in (Ndiaye et al., 2020). This is a detailed study enlisting recent developments in Healthcare IoT (HIoT). It also outlines the comparison between different implementation strategies for IoT systems before and during this pandemic. Golinelli et al. (2020) present a survey enlisting early efforts for the adoption of digital technologies in healthcare to fight against COVID-19, considering different categories such as diagnosis, prevention and surveillance. Chang (2020) emphasize the dire need to use AI techniques to combat future pandemics. They also highlight the limitations of the existing AI-based approaches towards the eradication of the pandemic. Finally, they conclude that there is a need to use data science in global health to produce better predictions that are helpful for policy makers. Abir et al. (2020) presented an analysis of how AI and IoT can potentially be used to fight against the COVID-19 pandemic.

7.3   IMPLICATIONS OF IOT FOR AUTONOMOUS MANUFACTURING

The IoT has been made possible by massive technological breakthroughs in computing power, miniaturization of wireless sensors, high-capacity networks, and big data analytics. Another important element is the fact that each of these technologies has come down so far in price, especially with the availability of cloud computing, that its use can spread widely. All of these technologies: computing power, miniaturized wireless sensors, high-capacity networks, and big data analytics, are already in use to some extent in manufacturing. Therefore, it makes sense that we are beginning to see an industrial application of the IoT. In fact, there is now a term for it, the Industrial Internet of Things (IIoT). The deployment of the IIoT in manufacturing plants is making it possible to extract much more information about the production process than has previously been possible, even though manufacturers have already been collecting a great deal of production data. The more extensive dataset made possible by the IIoT will ultimately move us toward autonomous production by making it possible to quickly update a production model to adapt to changing conditions, such as a custom order.

Compare this with the current practice in which a production system is designed and optimized to execute the exact same process repeatedly. In an autonomous production scenario, manufacturing systems will have the flexibility to adjust and optimize for each run of the task. Consider robotic operations as an example. Today, robots are programmed to perform specific tasks. As long as certain, pre-defined surrounding conditions are met, the robot will always execute the task identically. In a future with autonomous production, the robots will receive a task and will have to determine how to perform it in the most optimized way. Theoretically, for each run of the task, the work may be done differently. This will also hold true for a complete system. For instance, in the framing station of an automotive assembly plant, where major parts come together to form the complete body, each major component will be measured and the system will adjust itself to make the mating between the components optimal. Alternatively, each door panel opening will be measured and the best matching panel will be selected to the specific body (versus the current just-in-sequence method).

Industry 4.0 represents a significant opportunity for manufacturers to offer their customers enhanced product ranges and quality in tandem with more competitive

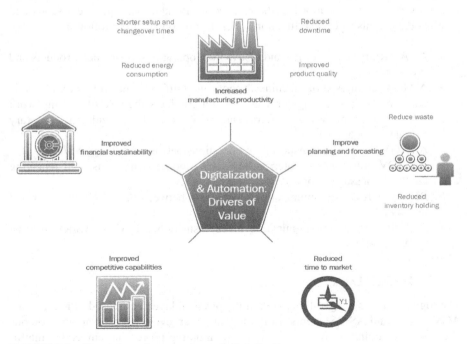

FIGURE 7.1 Key points of automation and digitalization.

pricing. Drawing on original research and a variety of published sources, this chapter summarizes the key points of value that digitalization and automation have been delivering to those pioneering companies that have already started to invest in the new wave of digitalized technology. These key points of value are summarized in the diagram below (Figure 7.1).

7.4   ARCHITECTURE OF SMART AUTONOMOUS MANUFACTURING SYSTEM USING IOT

The idea beyond our smart autonomous manufacturing is that CPSs monitor physical processes of autonomous manufacturing systems, create virtual copies of the physical world and make decentralized decisions. Over the IoT, smart autonomous manufacturing architecture (Figure 7.2) is built upon a physical layer, a communication layer and a logical layer. The functions of different layers will be described in the following subsections.

7.4.1   PHYSICAL LAYER

The physical layer implements processing and transportation operations on the shop floor, and is composed of Manufacturing Cells (MCs), Automated Guided Vehicles (AGVs), an Automated Storage/Retrieval System (AS/RS), Radio Frequency (RF) communication and Controller Area Network (CAN) communication. The code developed in this layer is in charge of moving, controlling and monitoring the equipment through RF and CAN communications in the real world. In addition, equipment sends and receives data at this level. The functions of each component are as follows:

- An AS/RS handles storage and distribution operations of finished products and raw material.
- A MC is composed of machines, robots and buffers. A machine provides several types of processing services; a robot conducts the role of shifting work pieces among machines, buffers, and AGVs; a buffer provides a temporary storage service to workpieces.
- An AGV deals with transportation operations between MCs and AS/RS.
- An AGV controller dispatches AGVs by sending commands, and receiving feedback messages from AGVs, simultaneously.
- RF is a wireless communication mechanism between the AGV controller and the AGVs.
- CAN is a wired communication mechanism between MCs, AS/RS and the AGV controller.

7.4.2   LOGICAL LAYER

The logical layer is a logical mapping of the physical layer, composed of MC agents, AGV agents and AS/RS agents. In the logical layer, every agent is an autonomous and cooperative unit, provided with strong computing power and embedded intelligent scheduling algorithms. Agents receive messages from entities (MCs and AGVs)

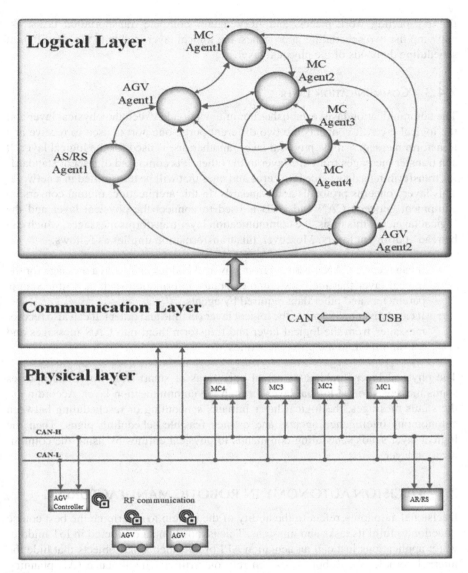

FIGURE 7.2 Smart autonomous manufacturing architecture.

in the physical layer and cooperate with each other to generate scheduling for the entities. The logical layer provides two types of scheduling services to the physical layer: scheduling of operations on machines and scheduling of transportation by AGVs.

For the scheduling of operations on machines, different MC agents produce feasible and optimized plans by competing processing tasks with each other. During the scheduling of transportation tasks, MC agents and the AS/RS agent compete AGVs for transporting work pieces, and AGV agents compete transportation tasks. By applying the two scheduling approaches, the logical layer is able to satisfy different scheduling demands of the physical layer.

7.4.3 COMMUNICATION LAYER

The communication layer establishes communication between the physical layer and the logical layer. It connects the two different ports: one port is used to receive and transform messages in the physical layer; another one is used for the logical layer. It can transfer messages from one layer to the other. It is composed of a set of standards for transforming a data value into zeros and ones that will be transmitted in a network. This layer concerns protocols and standards. In this architecture, mutual conversion equipment between CAN and USB is used to connect the physical layer and the logical layer. On this basis, the communication layer transforms messages, which can be read at different layers. Moreover, this transformation implies as follows:

- It can receive CAN messages from physical entities and build a message for the logical layer that includes complementary information such as a time stamp, parameters and other data required by agents.
- It can deliver messages to the logical layer (in time and order). It can also receive messages from the logical layer and transform them into CAN messages and transmit them to the physical entities.

The physical layer implements actual operations of smart shop floor, and updates status messages to the logical layer through the communication layer. According to the status messages, the logical layer initiates scheduling or rescheduling between autonomous intelligence agents, and creates feasible scheduling plans. Then the logical layer sends scheduling commands to physical entities by using the communication layer.

7.5 DECISION AUTONOMY IN ROBOTIC MANUFACTURING

Decisional autonomy refers to the ability of the system to determine the best course of action to fulfil its tasks and missions. This is mostly not considered in IoT middleware: applications just call an actuation API of so-called smart objects that hide the internal complexity. Roboticists often rely on Artificial Intelligence (AI) planning techniques based on predictive models of the environment and of possible actions. The quality of the plans critically depends on the quality of these models and of the estimate of the initial state. In this respect, the improved situational awareness that can be provided by an IoT environment can lead to better plans. Human-aware task planners use knowledge of the intentions of the humans inferred through an IoT environment to generate plans that respect constraints on human interaction. The IoT also widens the scope of decisional autonomy by making more actors and actions available, such as controllable elevators and doors. However, IoT devices may dynamically become available or unavailable, which challenges classical multi-agent planning approaches. A solution is to do planning in terms of abstract services, which are mapped to actual devices at runtime.

With the fourth industrial revolution, where robotics, CPS, and cloud technologies are merged, different domains are benefiting from rapid development. In this context, the IoRT systems can provide several advantages over traditional robotic applications, such as offloading computation-intensive tasks onto the cloud, accessing large amounts of data, and sharing information with other robots, aiming to learn new skills and knowledge from each other. Moreover, IoRT applications can also be used remotely, facilitating the work of both researchers and industrial operators, and making it more accessible, allowing cooperation between humans even from long distances.

Robot-based production represents the backbone of smart manufacturing, and the concept of industrial robots has been occurring as a continuous change over recent years, mainly due to the embedding of IoT technologies. The IoRT represents the main enabler of such change, through which manufacturing is embracing the concepts of Industry 4.0, embedding sensors, automation, and monitoring of products and processes. The Fourth Industrial Revolution has transformed how products are manufactured, adapting to such technological innovation, and aiming to produce high-quality goods and services. Smart manufacturing involves system flexibility, monitoring, and adaptation to change. Specifically, additive manufacturing is a critical process in terms of manufacturing methods. In fact, innovations in digital technologies that are occurring during the Fourth Industrial Revolution need to keep up with advancements in manufacturing processes and materials. In this scenario, with the aim of facilitating smart manufacturing by sensor systems, flexible electronics for additive manufacturing and their reliability during the processes are a critical consideration.

7.5.1 MULTI-LEVEL IoT-BASED MANUFACTURING CONTROL SYSTEM

Using the IoT, getting data is relatively straightforward, although it presents challenges in lower level services relating to hardware, low level embedded software, communication protocols and security that need to be addressed. The challenges in enhancing the efficiency of plant operations sit at a higher level because data coming from IoT devices lacks contextual information. Simply connecting lots of sensors and devices to a network is not enough, advanced analytics and dashboards must be built to deliver insights from the data generated. Without context, data collected from sensors around the plant cannot be translated into the information needed to make useful decisions both at a real-time, operational level and for analytical purposes such as big data, machine learning, predictive responses, and the like. By adding a Manufacturing Execution Systems (MES) layer, however, higher level contextual information collected from IoT devices can be stored and analyzed to make it relevant and useful. This data can be very diverse ranging across: quality data, status of machines and systems, faults and their causes, time and quantity reports, maintenance, yield, and so forth.

A manufacturing system can be composed of several integrated computers through a network for the purpose of control and supervision. The level of decision making is

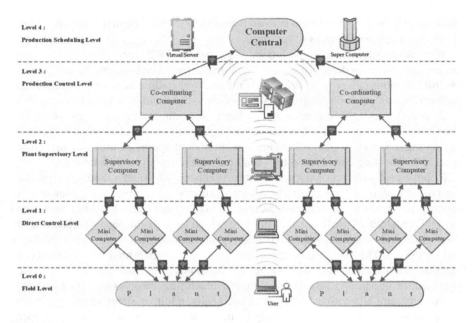

FIGURE 7.3   A multi-level computerized automation.

different in a plant. Thus, the data sharing and the required hardware are different. In Figure 7.3 a multi-level decision making and control structure for an advanced manufacturing system is shown.

Sensors measure temperature, humidity, air pressure and machine operating data in real time with full traceability from every step of the production process. Every asset on the plant floor is connected and every step in production in tracked in real-time. The MES layer becomes the performance management system. A MES can identify IoT devices, interact with them and control them. It can then be used to enforce production and quality processes to ensure that products are error free. In highly regulated environments, a MES can help companies to focus on quality and make compliance a natural result of excellent processes. Such processes may also extend into other systems such as Enterprise Resource Planning (ERP) or Product Lifecycle Management (PLM) or outside of the production facility, requiring wider integration of data across multiple plants and systems.

7.5.2   Modified SCADA Automation System

Supervisory Control and Data Acquisition (SCADA) is a system of software and hardware elements that allows industrial organizations to:

- Control industrial processes locally or at remote locations
- Monitor, gather, and process real-time data

- Directly interact with devices such as sensors, valves, pumps, motors, and more through Human-Machine Interface (HMI) software
- Record events into a log file.

SCADA systems are crucial for industrial organizations since they help to maintain efficiency, process data for smarter decisions, and communicate system issues to help mitigate downtime. The SCADA software processes, distributes, and displays the data, helping operators and other employees analyze the data and make important decisions.

The basic SCADA architecture begins with Programmable Logic Controllers (PLCs) or Remote Terminal Units (RTUs). PLCs and RTUs are microcomputers that communicate with an array of objects such as factory machines, HMIs, sensors, and end devices, and then route the information from those objects to computers with SCADA software.

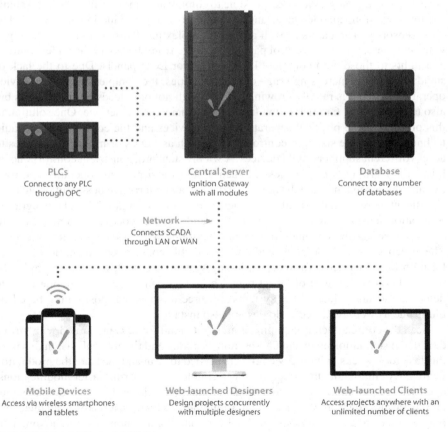

FIGURE 7.4   IoT-based SCADA.

SCADA systems are used by industrial enterprises to control and maintain efficiency, distribute data for smarter decisions, and communicate system issues to help mitigate downtime. SCADA systems work well in many different types of enterprise because they can range from simple configurations to large, complex installations.

The SCADA systems required for advanced and complex manufacturing systems employed today need to adopt networking systems. In the modified system, the communication between the system and the master station is done through the WAN protocols like the Internet Protocols (IP). Since the standard protocols used and the networked SCADA systems can be accessed through the internet, the vulnerability of the system is increased. However, the usage of security techniques and standard protocols means that security improvements can be applied in SCADA systems (Figure 7.4).

7.6 A PROPOSED IOT-BASED AUTONOMOUS MANUFACTURING SYSTEM

Using radio technologies it is also possible to employ new, mobile and flexible systems for the operation, maintenance, and diagnostics of the production facility. Today, most sensors and actuators as well as more complex mechatronic units are equipped with stationary, inflexible control panels that range from those with just a few buttons and lights to those with complete PC-based, color LCD panels. Due to the lack of standards and the increasing range of functionalities, the complexity of these device operating systems is rapidly growing, a fact which not only leads to higher costs but also to problems in familiarization training and maintenance service. One solution to this problem is the physical separation of the devices and the control panels. Radio technologies enable standard control devices such as PDA's or mobile telephones to access different suppliers' field devices. A widely standard, consistent control concept raises the learning conduciveness of such systems and prevents operational errors. Location independence and the advanced display and interactive possibilities enable a significant increase in the flexibility of device operations (Figure 7.5). The integration of location sensing systems with production and logistic processes is a major condition for meeting the demands for greater flexibility and shorter production cycles. The effective use of location data allows for flexible context-related applications and location-based services. Various positioning systems are deployed. For example, the floor is fitted with a grid of RFID tags. These tags can be read by mobile units to determine location data. Other systems for three-dimensional positioning based on ultrasonic as well as RF technologies are also installed.

Scan data of real factory equipment, in point cloud form, is integrated with virtual data in the simulation environment for more accurate and timely analysis. In this past, this task took weeks, with many engineers needed to scan and measure the production facility. By eliminating that step, the new technology will permit faster modifications, a critical requirement for autonomous production. Similarly event-driven simulation, (also known as discrete event simulation) which is already available, will play a bigger role as a fundamental tool for enabling autonomous production. This is because behind the flexibility and autonomy of this type of production scenario,

Implemented Technologies:
- Communication with Operator's Station via WLAN
- Decentralize process Control via RFID
- Wireless Sensor Network
- Parameterization with Universal Interaction Device
- Failure Indication via GPRS

Control Parameterization Set-up

FIGURE 7.5 Proposed IoT-based autonomous manufacturing.

there are very rigid rules that the system must adhere to. The self-driving vehicle is a good example. Without strict rules, these cars would wreak havoc. The challenge is to move from nominal planning to variable planning, where the result is driven by the varying surrounding conditions in the ecosystem. We already know that manufacturing is a key driver for economic growth, attracting investments, spurring innovation, and creating high-value jobs. All of the breakthrough and developments that are happening now – the building blocks for autonomous production – are already adding economic value. Imagine what will happen when autonomous production is routine. It is likely to be the revival that many predict for the manufacturing industry. We also see autonomous production as a way to address many global challenges such as a growing and aging population, climate change and resource scarcity.

Many opportunities exist for allocating production resources J and products P to multiple production cells and make possible a vast number of production system configurations with differing performance profiles. Hence, the Assembly System Configuration aims to explore a large variety of solutions, which can consist of one or more cells and fulfill design and performance requirements simultaneously. Therefore, the Assembly System Configuration automatically generates many configurations of the assembly system and analyzes their Key Performance Indicators (KPIs) to enable the exploration of the design space. To achieve this objective, multiple steps of design synthesis and analysis are executed in an automated way for each candidate solution.

7.6.1 Mathematical Formulation

Due to the evolution of the market requirements, in terms of product types to produce and their volumes, and the upgrading of the available assembly technologies in time, an assembly line design can easily become inappropriate and can require reconfiguration over time. Therefore, the assembly line design and management method must be able to cope with the evolution of requirements, also addressing how and when the assembly line configuration must change to match new production needs. All required dynamic data are sensed and collected via IoT technologies in the Industry 4.0 paradigm. To model the uncertain evolution of requirements, a probabilistic scenario model is proposed. A set of nodes O is defined, over a set T of periods. For each node, a probability of realization $P(O)$ is assigned at the beginning of the considered period (t_0). Each scenario node is characterized by a set of production requirements to be guaranteed if the realization of that specific scenario occurs by the IoT technology, leading to a modelling of the requirements over the time horizon $(t_0, t_1, t_2, \dots, T)$.

Based on detailed cell designs and the production parameters provided by the layout configuration module, the production planning and control module is responsible for testing the robustness of the designed system under specific due-dates imposed by the customers (see Figure 2). The first production planning activity optimizes the production schedule and the lot sizes for user defined due-time performance. Besides, a control tool evaluates the defined system configuration under the specific schedule, considering the effects of stochastic parameters and random events on logistics-related performance indicators. The input of the production planning activity are the set of products that are assembled in the system, the number of available resources, the detailed layout of the system as well as the due dates coming from the customers. Due dates can be predicted in the early system configuration stage by knowing contractual delivery frequency requested by the customer, and they have significant impact on the applied production lot-sizes and, therefore, the operational costs. The control tool is directly linked to the production planning activity, as the main inputs of the analysis are the calculated production plan, the system configuration with detailed data of the processes, as well as logistics related data, for example, actual inventory and backlog levels. Production planning is done on a discrete time horizon W and the resolution of the plan is a working shift (w). The objective is to calculate the production lots $x_{p,w,c}$ respecting the available capacities, cycle (t^m_p) and setup (t^s) time constraints. In the model, setups are expressed with the binary variables $z_{p,w,c}$ and $y_{p,w,c}$. When assembling a certain product type, a definite amount of resources $r_{j,p}$ is required, and a given amount n_j of resource type j is available for use at the beginning of the period. The order demands d_p need to be satisfied by delivering certain amount $s_{p,w}$ of products to customers. In the production planning, holding inventory of products ($i_{p,w}$) is allowed, however, it has certain costs c^i. Similarly, planned backlogs ($b_{p,w}$) might occur, but they are also penalized with cost c^b per product and shift. The objective (1) of the problem is to minimize the total backlog and inventory costs that incur in the period. Production planning is formulated as an integer programming problem:

$$\min \sum_p \sum_w \left(c^b b_{p,w} + c^i i_{p,w} \right). \tag{1}$$

Such that:

$$\sum_c \sum_p r_{j,p} y_{p,w,c} \le n_j, \qquad \forall w, j, \tag{2}$$

$$\sum_p \left(t_p^m x_{p,w,c} + t^s z_{p,w,c} \right) \le t^p, \qquad \forall w, c, \tag{3}$$

$$d_p \le s_{p,w}, \qquad \forall p, w, \tag{4}$$

$$i_{p,w} - b_{p,w} = i_{p,w-1,c} - s_{p,w} + \sum_c x_{p,w,c}, \qquad \forall w, p. \tag{5}$$

$$x_{p,w,c} \ge 0 \text{ and } Integer; z_{p,w,c} \text{ and } y_{p,w,c} \in \{0.1\} \tag{6}$$

The first constraints include the limited amount of resources (2) and human capacities (3). Inequality (4) states that demands must be fulfilled, and the balance equation (5) links the subsequent production shifts. For the calculation of the setups ($z_{p,w,c}$ and $y_{p,w,c}$), the multi-item single-level lot sizing model is applied (LS-C-B/M1). The cell-product assignments ($a_{p,c}$, equals 1 if product p is assigned to cell c, 0 otherwise) are determined by the previous modules, however, the assignment of resources to cells needs to be optimized by the production planning module, in order to avoid conflicts. Relations (6) show the types and signs of the decision variables.

7.6.2 IMPLEMENTATION STUDY

The proposed approach has been applied for each of the considered scenario nodes. Firstly, the design synthesis module generates design candidates according to different production strategies and analyses their performances.

The results of the whole approach applied on scenario path o_0 -o_{1A} -o_{2B} are reported in Table 7.1. The first row refers to the robust solution. The second row refers to the optimal solution for the considered scenario path only, obtained by choosing the best configuration solution at each step. The last row reports the solution in which an optimal solution for o_0 is used in every time bucket. The solutions are compared in terms of purchasing, reconfiguration, storage and operational costs. Results demonstrate that the robust solution ensures a lower total discounted cost compared to the optimal solution for the single scenario path (771 730 unit of money against 806 909 unit of money), the difference is mainly due to the fact that the robust solution behaves proactively, purchasing additional pieces of equipment in advance, while the other solution has to react to the changes through a reconfiguration step whose impact on the cost is relevant (10 000 unit of money).

TABLE 7.1
Numerical results for the industrial case

	Cost type	t_1	t_2	t_3	Total
Robust approach (overall approach)	Functional assembly purchases	358,883	0	0	358,883
	Module purchases	50,000	0	0	50,000
	Reconfiguration	0	0	0	–
	Storage	0	12,000	0	12.000
	Operative	92,010	106,002	78,894	276,906
	Tool purchases	45,000	20,000	20,000	85,000
	Total (discount)	545,893	133,412	92,425	771,730
Single path optimum (best configuration is chosen for each scenario)	Functional assembly purchases	358,883	0	0	358,883
	Module purchases	40,000	0	10,000	50,000
	Reconfiguration	0	10,000	10,000	20,000
	Storage	0	18,000	0	18,000
	Operative	100,776	103,542	83,850	288,168
	Tool purchases	45,000	20,000	20,000	85,000
	Total (discount)	544,659	146,502	115,748	806,909
Single node optimum (best o_0 configuration is used in every scenario)	Functional assembly purchases	358,883	0	I.S	358,883
	Module purchases	40,000	0	I.S	40,000
	Reconfiguration	0	10,000	I.S	10,000
	Storage	0	18,000	I.S	18,000
	Operative	100,776	103,542	I.S	204,318
	Tool purchases	45,000	20,000	I.S	65,000
	Total (discount)	544,659	146,502	-	691,161

I.S: Infeasible solution

7.7 COMPREHENSIVE IOT-ROBOTIC MANUFACTURING SYSTEM IN PANDEMICS

Application of integrated robotics and the IoT leads to enhancing the technologies and opportunities related to healthcare specifically in facing pandemics such as the novel Coronavirus. The major infection route of pandemics is contact between people while IoRT provides a contactless technology to handle production tasks (or even in other areas such as service, agriculture, medical treatments, and so forth.) leading to a safer environment. Meanwhile, production tasks are not stopped anymore while the whole process is performed and monitored by intelligent robots (RajKumar et al., 2019).

The COVID-19 pandemic implies the vulnerability of workers to infection causing tremendous loss in production, services, customer relationship management and all other enterprise functional modules. IoRT could handle all manufacturing tasks, perform measurement, and collect data and information sensed by the sensors to facilitate intelligent decision-making.

IoRT includes triple intelligent modules:

1   Sensors that are used by robots to sense their environment to monitor the manufacturing floor and transfer data with other intelligent devices
2   An analytical core which is able to mine the data collected and process it using computational capabilities like edge computing (local computation) rather than cloud computing, thus preventing huge amount of data transfer
3   Intelligent action, that is, a robot can act based on the collected data and analysis leading to a complete cyber-physical process where cooperation between machine2machine and man2machine gives rise to the emergence of new services like IoRT-maintenance, IoRT-quality control, IoRT-inventory control and the like.

Robots are appropriate substitutes for the human workforce with respect to increases in labor price. Artificial intelligence is a plus in robotic systems to keep a standard production mode and required quality during the production cycle time. The process of man2machine boosts productivity with a smaller number of workers.

The IoRT is capable of facing pandemics, keeping a certain standard of manufacturing requirements and on the other hand providing remote man2machine communication through the internet for control and coordination purposes, promising appropriate multi–dimension data analysis and decision-making. (See Figure 7.6).

It has been over three years since the global COVID-19 pandemic has had widespread effects on society, including rapid technological developments. During the pandemic, we saw the adoption of increased automation procedures and technologies to reduce the amount of interpersonal physical interaction. E-commerce thrived like never before. Meanwhile, cybersecurity has become an issue of paramount importance in the face of several large-scale cybercrime attacks.

When it comes to the Internet of Things, or IoT, the coronavirus sparked new uses for pre-existing technology. Drones were deployed to deliver medical supplies in China, while governments utilized automatic sensing devices and Bluetooth-enabled gadgets to track potential contagions.

Now that the pandemic has subsided in intensity, we are entering the next phase of society: the post-pandemic era. Incorporating both permanent and temporary shifts concerning the pandemic, our societal approach to the Internet of Things looks different now than it did before 2020. In this article, we will take a look at how the IoT has continued to evolve and find its place in the market since the pandemic. We will focus on specific IoT innovations that are set for success now, and we will examine the potential cybersecurity risks of contemporary IoT devices and gadgets.

The sector with the greatest market potential for IoT development is manufacturing. IoT networks and devices will be used to make manufacturing processes more efficient, optimize communications between employees and management, and streamline asset management in day-to-day operations.

Since the IoT facilitates seamless remote interactions, manufacturing facilities can now include employees and assets spread across a broad geographical range.

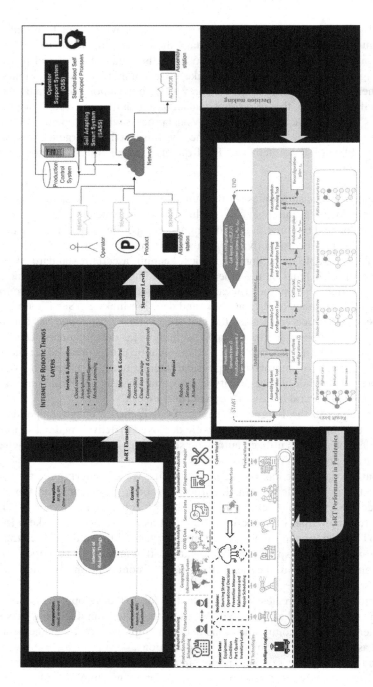

FIGURE 7.6 Comprehensive IoT-Robotic Manufacturing System in Pandemics.

The market for IoT solutions is proving most viable in factory settings, according to expert predictions. In the manufacturing field, IoT networks can be utilized to help combat ongoing supply chain issues that arose during the pandemic. The rise of the industrial IoT, however, has also made it an increasingly attractive target for cybercriminals. This is why securing industrial IoT devices will also need to become a bigger priority for businesses as well.

7.8 DISCUSSION

Even with the growing proliferation of the IIoT, we are not at the point where autonomous production is widespread or routine, although there are some plants implementing some elements of it today. What is already widespread and routine, however, is the solid foundation for autonomous production, which can be seen in practice in the many plants that have adopted digital manufacturing. Digital manufacturing solutions provide manufacturers with vast capabilities for designing and evaluating their processes virtually. In a digital manufacturing environment such as our company's manufacturing planning and management solutions, the physical world is replicated in a model-driven database. Digital tools and methods are used to design the physical manufacturing system, including its logical controls. The result is a comprehensive virtual model of the manufacturing process that crosses multiple engineering disciplines such as tooling, process, logistics, quality and product. Digital simulation tools make it possible to validate and optimize the processes, tools and the control algorithms, and the interactions between them, all in a virtual environment prior to commissioning the system on the shop floor. Beyond digital manufacturing is the digital factory, which is comprised of additional technological layers. A digital factory requires an infrastructure to connect devices, the ability to identify where connectivity legitimately adds value and is not merely intrusive, and software platforms that will unlock the torrent of data.

To put the Digital Factory in its proper place on the road to autonomous production, we need to look first at how it can improve manufacturing flexibility, since that is a critical element of autonomous production. Traditional production uses a sequential process flow between production modules (a moving line), where each module has a dedicated task to perform in a given sequence.

A flexible process flow between production modules, made possible by Digital Factory solutions, allows different modules to be configured for each production instance. Such a production system is more resilient to changes, and allows greater variation in manufacturing scenarios (production mix, production volume) and product range. Digital Factory solutions optimize production asset utilization by constantly monitoring, controlling and analyzing the production ecosystem and making online decisions. They can do this for physical assets, such as machines, inventory or energy consumption, such as time to completion. Additional technologies are still needed before autonomous production is a reality, but they are arriving on a regular basis. For example, Siemens PLM Software and our partner, Bentley Systems, recently introduced a new point cloud technology that makes it possible to capture the exact position of the production and logistical assets on the shop floor, providing in almost real-time the actual conditions of the shop floor.

7.9 CONCLUSIONS

The IoT needs both data collection and actuation features but it also needs contextual information and orchestration to make this data useful and enable automation scenarios to be built around it. Although the IoT can be integrated into shop-floor systems relatively easily, having an MES that supports the necessary decentralized logic of the smart shop-floor will enable automated production of customized products. With the emerging information technologies, such as the IoT, big data, and cloud computing together with artificial intelligence technologies, we believe the smart factory of Industry 4.0 can be implemented. The smart machines and products can communicate and negotiate with each other to reconfigure themselves for flexible production of multiple types of products. The massive data can be collected from smart artifacts and transferred to the cloud. This enables the system-wide feedback and coordination based on big data analytics to optimize system performance. The above self-organized reconfiguration and big data-based feedback and coordination defines the framework and operational mechanism of the smart factory. The smart factory helps to implement the sustainable production mode to cope with global challenges. It can lead to novel business modes and even affect our lifestyle. Although the implementation of smart factory is still facing some technical challenges, we are walking on the right path by simultaneously applying the existing technologies and promoting technical advancements. With the existing technologies, some application demonstrations have already been built. Therefore, the smart factory and Industry 4.0 can be implemented in a progressive way, along with the unstoppable technical advancements. In the future, we will continue to develop our prototype design and focus on the key enabling technologies.

The IoRT emerges as a novel paradigm in modern manufacturing systems. Cooperative learning based on the sensed data provides a collective knowledge being employed by robots to handle their allocated tasks in an intelligent manner. IoT sensors help in collecting data and transmitting it through a safe and fast route for further control or decision making tasks. Motion path planning and other controlling behaviors and actions of robots are based on the data gathered by sensors. The IoT focuses on pervasive sensors while robots uses interactive ones to handle automated production and execute production tasks in any situation. Robots can perform their tasks in pandemics where the human work force are not able to cooperate. This capability provide a huge opportunity for producers to absorb a larger amount of the market share in competition.

BIBLIOGRAPHY

Abir, S., Islam, S.N., Anwar, A., et al. (2020). Building Resilience against COVID-19 pandemic using artificial intelligence, machine learning, and IoT: A survey of recent progress. *Internet of Things*, 1, 506–528.

Alam, T. (2020). Internet of Things and Blockchain-based framework for Coronavirus (Covid-19) disease. *Social Science Research Network*.

Alcácer, V., & Cruz-Machado, V. (2019). Scanning the Industry 4.0: A literature review on technologies for manufacturing systems. *International Journal of Engineering, Science and Technology*, 22, 899–919.

Bai, L., Yang, D., Wang, X., et al. (2020). Chinese experts' consensus on the Internet of Things-aided diagnosis and treatment of coronavirus disease 2019 (COVID-19). *Clinical eHealth*, 3, 7–15.

Brem, A., Viardot, E., & Nylund, P.A. (2021). Implications of the coronavirus (COVID-19) outbreak for innovation: Which technologies will improve our lives? *Technological Forecasting and Social Change*, 163, 120451.

Burnwal, S., & Deb, S., (2013). Scheduling optimization of flexible manufacturing system using cuckoo search-based approach. *The International Journal of Advanced Manufacturing Technology*, 64(5–8), 951–959.

Chamola, V., Hassija, V., Gupta, V., et al. (2020). A comprehensive review of the COVID-19 pandemic and the role of IoT, drones, AI, Blockchain, and 5G in managing its impact. *IEEE Access*, 8, 90225–90265.

Chang, A.C. (2020). Artificial Intelligence and COVID-19: Present state and future vision. *Intelligence-Based Medicine*, 3, 100012.

Dong, Y., & Yao, Y.D. (2020). IoT Platform for COVID-19 prevention and control: A survey. *arXiv*. arXiv:2010.08056

Elansary, I., Darwish, A., & Hassanien, A.E. (2021). The future scope of internet of things for monitoring and prediction of COVID-19 Patients. In *Digital Transformation and Emerging Technologies for Fighting COVID-19 Pandemic: Innovative Approaches*. Springer, Cham, Switzerland, pp. 235–247.

Erol, R., Sahin, C., Baykasoglu, A., et al. (2012). A multi-agent based approach to dynamic scheduling of machines and automated guided vehicles in manufacturing systems. *Applied Soft Computing*, 12(6), 1720–1732.

Fazlollahtabar, H. (2016). Parallel autonomous guided vehicle assembly line for a semi-continuous manufacturing system. *Assembly Automation*, 36(3), 262–273.

Fazlollahtabar, H. (2018a). Lagrangian relaxation method for optimizing delay of multiple autonomous guided vehicles. *Transportation Letters*, 10(6), 354–360.

Fazlollahtabar, H. (2018b). Scheduling of multiple autonomous guided vehicles for an assembly line using minimum cost network flow. *Journal of Optimization in Industrial Engineering*, 11(1), 185–193.

Fazlollahtabar, H. (2019a). An effective mathematical programming model for production of automatic robot path planning. *The Open Transportation Journal*, 13(1), 11–16.

Fazlollahtabar, H. (2019b). Triple state reliability measurement for a complex autonomous robot system based on extended triangular distribution. *Measurement*, 139, 122–126.

Fazlollahtabar, H. (2020). Comparative simulation study for configuring turning point in multiple robot path planning: Robust data envelopment analysis. *Robotica*, 38(5), 925–939.

Fazlollahtabar, H. (2021). Robotic manufacturing systems using Internet of Things: New era of facing pandemics. *Automation, Robotics & Communications for Industry*, 4.0, 82.

Fazlollahtabar, H. (2022). Internet of Things-based SCADA system for configuring/reconfiguring an autonomous assembly process. *Robotica*, 40(3), 672–689.

Fazlollahtabar, H., & Hassanli, S. (2018). Hybrid cost and time path planning for multiple autonomous guided vehicles. *Applied Intelligence*, 48, 482–498.

Fazlollahtabar, H., & Jalali, S.G. (2013). Adapted Markovian model to control reliability assessment in multiple AGV. *Scientia Iranica*, 20(6), 2224–2237.

Fazlollahtabar, H., Mahdavi-Amiri, N., & Muhammadzadeh, A. (2015). A genetic optimization algorithm for nonlinear stochastic programs in an automated manufacturing system. *Journal of Intelligent & Fuzzy Systems*, 28(3), 1461–1475.

Fazlollahtabar, H., & Niaki, S.T.A. (2017a). Binary state reliability computation for a complex system based on extended Bernoulli trials: Multiple autonomous robots. *Quality and Reliability Engineering International*, 33(8), 1709–1718.

Fazlollahtabar, H., & Niaki, S.T.A. (2017b). Integration of fault tree analysis, reliability block diagram and hazard decision tree for industrial robot reliability evaluation. *Industrial Robot: An International Journal*, 44(6), 754–764.

Fazlollahtabar, H., & Niaki, S.T.A. (2017c). *Reliability Models of Complex Systems for Robots and Automation.* CRC Press.

Fazlollahtabar, H., & Niaki, S.T.A. (2018a). Cold standby renewal process integrated with environmental factor effects for reliability evaluation of multiple autonomous robot system. *International Journal of Quality & Reliability Management,* 35(10), 2450–2464.

Fazlollahtabar, H., & Niaki, S.T.A. (2018b). Modified branching process for the reliability analysis of complex systems: Multiple-robot systems. *Communications in Statistics-Theory and Methods,* 47(7), 1641–1652.

Fazlollahtabar, H., & Saidi-Mehrabad, M. (2015a). *Autonomous Guided Vehicles: Methods and Models for Optimal Path Planning.* Springer International Publishing, Germany.

Fazlollahtabar, H., & Saidi-Mehrabad, M. (2015b). Risk assessment for multiple automated guided vehicle manufacturing network. *Robotics and Autonomous Systems,* 74, 175–183.

Fazlollahtabar, H., & Saidi-Mehrabad, M. (2019). *Cost Engineering and Pricing in Autonomous Manufacturing Systems.* Emerald Publishing Limited.

Fazlollahtabar, H., Saidi-Mehrabad, M., & Balakrishnan, J. (2015a). Mathematical optimization for earliness/tardiness minimization in a multiple automated guided vehicle manufacturing system via integrated heuristic algorithms. *Robotics and Autonomous Systems,* 72, 131–138.

Fazlollahtabar, H., Saidi-Mehrabad, M., & Balakrishnan, J. (2015b). Integrated Markov-neural reliability computation method: A case for multiple automated guided vehicle system. *Reliability Engineering & System Safety,* 135, 34–44.

Fazlollahtabar, H., Saidi-Mehrabad, M., & Masehian, E. (2015). Mathematical model for deadlock resolution in multiple AGV scheduling and routing network: A case study. *Industrial Robot: An International Journal,* 42(3), 252–263.

Fazlollahtabar, H., Saidi-Mehrabad, M., & Masehian, E. (2021). Robotic industrial automation simulation-optimization for resolving conflict and deadlock. Assembly Automation, 41(4), 477–485.

Fazlollahtabar, H., & Shafieian, S.H. (2014). An optimal path in an AGV-based manufacturing system with intelligent agents. *Journal for Manufacturing Science and Production,* 14(2), 87–102.

Gen, M., Lin, L., (2014). Multiobjective evolutionary algorithm for manufacturing scheduling problems: State-of-the-art survey. *Journal of Intelligent Manufacturing,* 25(5), 849–866.

Golinelli, D., Boetto, E., Carullo, G., et al. (2020). Adoption of digital technologies in health care during the COVID-19 pandemic. *Systematic Review of Early Scientific Literature. Journal of Medical Internet Research,* 22, e22280.

Javaid, M., Haleem, A., Vaishya, R., et al (2020). Industry 4.0 technologies and their applications in fighting COVID-19 pandemic. *Diabetes & Metabolic Syndrome: Clinical Research & Reviews, 14,* 419–422.

Ju, C., & Son, H.I. (2020). Modeling and control of heterogeneous agricultural field robots based on Ramadge–Wonham theory. *IEEE Robotics and Automatic Letters,* 5, 48–55.

Kalhori, S.R.N., Bahaadinbeigy, K., Deldar, K., et al. (2021). Digital health solutions to control the COVID-19 pandemic in countries with high disease prevalence: Literature review. *Journal of Medical Internet Research,* 23, e19473.

Kelly, J.T., Campbell, K.L., Gong, E., et al. (2020). The Internet of Things: Impact and implications for health care delivery. *Journal of Medical Internet Research,* 22, e20135.

Li, B. H., Zhang, L., Wang, S. L., et al. 2010, Cloud manufacturing: A new service oriented networked manufacturing model. *Computer Integrated Manufacturing Systems,* 16(1), 1–7.

Mahbub, M. (2020). A smart farming concept based on smart embedded electronics, internet of things and wireless sensor network. *Internet Things*, 9, 100161.

Mokbel, M.F., Abbar, S., & Stanojevic, R. (2020). Contact tracing: Beyond the Apps. *arXiv*, arXiv:2006.04585.

Mourtzis, D., Vlachou, E., & Milas, N., (2016). Industrial big data as a result of IoT adoption in manufacturing. *Procedia CIRP*, 55, 290–295.

Nasajpour, M., Pouriyeh, S., Parizi, R.M., et al. (2020). Internet of Things for current COVID-19 and future pandemics: An exploratory study. *arXiv*, arXiv:2007.11147.

Nayak, J., Naik, B., Dinesh, P., et al. (2021). Intelligent system for COVID-19 prognosis: A state-of-the-art survey. *Applied Artificial Intelligence*, 51, 2908–2938.

Ndiaye, M., Oyewobi, S.S., Abu-Mahfouz, A.M., et al. (2020). IoT in the Wake of COVID-19: A survey on contributions, challenges and evolution. *IEEE Access*, 8, 186821–186839.

Ngai, E.W.T., Chau, D.C.K., Poon, J.K.L., et al. (2012). Implementing an RFID based manufacturing process management system: Lessons learned and success factors. *Journal of Engineering and Technology Management*, 29(1), 112–130.

Njike, A. N., Pellerin, R., & Kenne, J. P., (2012). Simultaneous control of maintenance and production rates of a manufacturing system with defective products, *Journal of Intelligent Manufacturing*, 23(2), 323–332.

Ounnar, F., & Pujo, P. (2012). Pull control for Job Shop: Holonic Manufacturing System approach using multicriteria decision-making. *Journal of Intelligent Manufacturing*, 23(1), 141–153.

Rahman, M.S., Peeri, N.C., Shrestha, N., et al. (2020). Defending against the Novel Coronavirus (COVID-19) Outbreak: How Can the Internet of Things (IoT) help to save the World? *Health Policy Technology*.

RajKumar, K., Kumar, C.S., Yuvashree, C., et al. (2019). Portable surveillance robot using IoT. *International Research Journal of Engineering and Technology*. (IRJET), 6, 94–97.

Rymaszewska, A., Helo, P., & Gunasekaran, A., (2017). IoT powered servitization of manufacturing–an exploratory case study. *International Journal of Production Economics*, 192, 92–105.

Shi, J., Wan, J., Yan, H., et al. 2011, *A Survey of Cyber-Physical Systems, International Conference on Wireless Communications and Signal Processing (WCSP)*, IEEE, 1–6.

Shojaeifar, A., Fazlollahtabar, H., & Mahdavi, I. (2016). Decomposition versus minimal path and cuts methods for reliability evaluation of an advanced robotic production system. *Journal of Automation Mobile Robotics and Intelligent Systems*, 10(3), 52–57.

Singh, R.P., Javaid, M., Haleem, A., et al. (2020). Internet of things (IoT) applications to fight against COVID-19 pandemic. *Diabetes Metab. Syndr. Clin. Res. Rev.*, 14, 521–524.

Tao, F., Cheng, Y., Da, Xu. L., et al. (2014). CCIoT-CMfg: Cloud computing and Internet of things-based cloud manufacturing service system. *IEEE Transactions on Industrial Informatics*, 10(2), 1435–1442.

Thramboulidis, K., & Christoulakis, F. (2016). UML4IoT—A UML-based approach to exploit IoT in cyber-physical manufacturing systems. *Computers in Industry*, 82, 259–272.

Ting, D.S.W., Carin, L., Dzau, V., et al. (2020). Digital technology and COVID-19. *Nature Medicine*, 26, 459–461.

Vaishya, R., Javaid, M., Khan, I.H., et al. (2020). Artificial Intelligence (AI) applications for COVID-19 pandemic. *Diabetes Metab. Syndr. Clin. Res. Rev.*, 14, 337–339.

Valecce, G., Micoli, G., Boccadoro, P., et al. (2019). Robotic–aided IoT: Automated deployment of a 6TiSCH network using an UGV. *IET Wireless Sensor Systems*, 9, 438–446.

8 Sustainability Evaluation in Pandemics

8.1 INTRODUCTION

The 17 Sustainable Development Goals are considered an appeal for action by all countries aiming for sustainable development while caring for the environment and the well-being of all inhabitants of our planet. Sustainable Consumption and Production (SPC) is embedded in the SDGs; in fact, sustainability and consumption are at the core of sustainable development (Sala and Castellani, 2019), which aims to support a modification to sustainable patterns of production and consumption, as proposed by Goal 12, that clearly refers to the need to ensure sustainable consumption and production patterns (UNDP, 2016). Some of those patterns are related to some aspects of consumption, such as food waste. It is supposed that every year about one-third of all produced food ends up decomposing in the bins of consumers and sellers. Another problem has to do with the excessive (and non-efficient) household consumption of energy and the generation of CO_2 emissions (Travassos et al., 2020). The wasting of water and water pollution are other issues caused by unsustainable patterns of production and consumption.

In general, business, and economic ecosystems are characterized by a certain degree of vulnerability. This vulnerability is usually due to shocks, such as an economic crisis, natural disasters, and wars and, more recently, biological risks such as COVID-19 (Arcese et al., 2020). Small and medium sized business activities (SME) are present in different commercial sectors and can be generally found in small cities, small towns, and municipalities. Currently, in Italy but also in the rest of Europe, they are characterized by negative economic cycles, strongly affected by the pandemic condition (Bellandi, 2020).

The upward trend of urbanization characterized by a rapidly growing number of city dwellers (Goddard,2021) and an increasing population density is exacerbating sustainability issues such as air and water pollution, insufficient public waste management and dependence on non-renewable energy sources; but also issues the quality of citizens' lives such as congestion, inadequate development of public transport systems and the lack of digitalization of public services (Arcese, et al., 2019). The current model of city development is in contrast with what today are the main forms of fighting the pandemic, such as social distancing, isolation, and individual forms of work such as smart working at home.

DOI: 10.1201/9781003400585-8

One method to assess the sustainability performance of products and/or services is Life Cycle Sustainability Assessment (LCSA), which assesses product performance considering the environmental, economic, and social dimensions over the life cycle (Finkbeiner et al., 2010; Traverso et al., 2012). Even if the LCSA methodology was defined about ten years ago, it has not been standardized yet. Moreover, this methodology and its indicators do not assess resilience, the importance of which has increased when considering climate change and pandemic phenomena such as COVID-19. Resilience in sustainability assessments and in the Life Cycle Sustainability Assessment context will be proposed.

The COVID-19 pandemic has decelerated global progress toward achieving the Sustainable Development Goals (SDGs). These aspects correspond to the first three goals in the world sustainable development agenda 2030: poverty alleviation (SDG1), hunger abatement (SDG2), and healthcare promotion (SDG3). The environmental and economic impacts of the epidemic are tightly linked to climate change mitigation (SDG13) and sustainable economic growth (SDG8).

8.2 RELATED WORKS

The United Nations (UN) introduced SDGs on 25 September 2015. These goals will be accomplished over fifteen years, beginning 1 January 2016 and ending 31 December 2030. The 2030 Agenda is a collection of 17 SDGs with 169 objectives designed to achieve a more sustainable, resilient, and prosperous world. COVID-19 represents a major threat to achieving SDGs by 2030. Besides, it endangers the growth prospects of developing countries that lack the expertise or resources required to deal with the associated economic and social problems. This situation is considered to be a severe setback for the primary objectives of sustainable development, which are inclusion and no one being left behind. Global unity and shared commitment must be critical in rebuilding momentum toward attaining the 2030 Agenda for Sustainable Development. The UN has announced a strategy to "defeat the virus and establish a better world" to confront this pandemic. The plan asks for the world's most powerful countries to establish decisive policy action and provide financial and technical assistance to the poorest and most vulnerable people. Besides, the UN has started a fundraising campaign to gather USD 2 billion to combat COVID-19.

It is hard to find consensus regarding the definition of sustainable consumption. Mont and Plepys (2008) point out that some authors deal with the topic as a production issue and recommend improvements/efficiency in the production processes to reduce the environmental problems. Others tend to associate sustainable consumption with the greening of markets, or the change to simplified lifestyles. Adopting a sustainable lifestyle is an option that consumers can take or not. It is possible to guide and educate them to follow a certain consumption pattern, but most of the time it is not possible to force them to adopt sustainable consumption actions (Marchand and Walker, 2008. Black and Cherrier (2010) considered that it is a practice in which individuals manifest their concern with the sustainability of the planet and are willing to spend time and financial resources to follow their cause. Thus, sustainable consumption is considered an umbrella concept (Paço and Laurett, 2019).

Nhamo and Ndlela (2021) corroborate the aforementioned statement and argue that lifestyle and psychological distress resulting from the pandemic caused profound changes in consumption behavior in homes. Barnes et al. (2020) commented that in situations, such as the COVID-19 pandemic, consumers may assume a panic behavior and acquire items impulsively due to uncertainties. During the COVID-19 pandemic, the social isolation associated with the evolution in digital purchasing technologies stimulated unnecessary consumption, going against what sustainable consumption preaches. For Sheth (2020), the pandemic's consequences on technology's evolution and the intersection between professional and personal life will consolidate new habits. It is worth remembering that the United Nations has disseminated goals related to sustainable consumption since 2015, via SDG 12 (United Nations – Department of Economic and Social Affairs, 2020). Responsible sustainable consumption will reduce the pressure on the planet's natural resources. Today, the ecological footprint indicates a deficit when comparing the consumption pace of the world population with the Earth's regenerative bio capacity (WWF). Severo et al. (2020) performed a study to better understand the COVID-19 pandemic's impacts on social responsibility, sustainable consumption and environmental awareness in different generations of Brazilians and Portuguese. In general, the results showed that the pandemic's consequences were characterized as an important element in changing people's habits towards becoming more sustainable.

Cavallo et al. (2020) carried out a survey with Italian citizens on consumption habits during the COVID-19 pandemic. The authors noted an increase in the purchase of essential items, greater concern with food security, more time preparing meals at home and a greater increase in home deliveries. Of course, all these changes have negative and positive points regarding sustainability, but in general, the authors consider the balance to be positive and envisage changes in consumption patterns in the coming years. A similar study was carried out by Marty et al. when analyzing the food habits of French citizens during the pandemic. The authors noted changes associated with health, sustainability and ethical behavior.

The search for a link between the concept of resilience and that of sustainability turns out to be an extremely attractive research topic and not a new exploration. As a basis of this research, there is the Ecology and the Industrial Ecology (as consequence) because the sustainability and the resilience concepts historically are based on a quantitative and engineering approach. From the point of view extrapolated from the Holling theories, the system is based on conditions of equilibrium, and we analyze how quickly the system, after a disturbance, returns to equilibrium. Conversely, many ecological and socio-ecological systems may spend more time out of equilibrium and in several possible states ("stable alternative"). The flexibility of the concept manifests a complexity of analysis and the often-changing nature of their context. Indeed, keeping context in mind is of prime importance in analyzing how close resilience and sustainability are (Walker et al., 2004). This is done mainly because the systems (industrial, social and environmental) are considered to be connected to each other. Talking today, therefore, of a resilience model with a single technical approach would be reductive. In fact, if we consider all the actions implemented in recent years to reduce environmental impacts and the consumption of resources, we realize that efforts are often in vain and global unsustainability does not stop. The ecosystem is a complexity of variables that include political choices, business strategies and the introduction of new technologies.

Resilience is a property of systems essential for sustainability (Darestani et al., 2021; Pizzol, 2015). Indeed, according to several authors (D'adamo and Rosa, 2020; Sarkis, 2020) resilience and sustainability are related concepts. Several research projects have identified cases where resilience plays a key role for sustainability as well as in climate change adaptation in cities (Hunt and Watkiss 2011), water management (Pizzol et al., 2013), urban management (Zhang et al., 2012), resource management, energy security of gas supply (Scotti and Vedres 2012), distributed economies (Johansson et al., 2005; Mirata et al., 2005), and so forth.

Due to Covid-19 and to pursue the path to sustainability, the industrial sector must assess the risks in product development and production processes by implementing a resilient system (Diaz-Elsayed et al., 2020). Barbosa (2021) has identified this resilience as a new research opportunity, given the need to create a resilient ecosystem in order to re-establish a link with consumers. It is essential to build resilience through risk-informed sustainable development to recovery from Covid-19, which aims to generate sustainable and resilient communities. The Covid-19 outbreak represents an opportunity to implement sustainable practices reducing and managing several risks. In addition, Pizzol et al. (2015) show that increasing product systems' resilience does not necessarily lead to a decrease in their ecoefficiency. However, the issue of resilience is not explored in Life Cycle Assessment (LCA) studies. a Few researchers have mentioned LCA in connection with design for resilience (Fiksel 2003; Korhonen and Seager 2008; Mu et al., 2011; Seager 2008; Kou and Zhao 2011).

Companies to pursue the trend of sustainability should design solutions that take into account the impacts of climate change and have the ability to adapt to changing conditions after COVID-19 (O'Connell and Hou, 2015). For instance, the existing literature highlights that enhancing resilience to global environmental change is an emerging research trend as it is a theme that remains unexplored (Simon, 2015; Worthy et al., 2015).

Since 2000, Life Cycle Management (LCM) has considered only the environmental impacts, and it is linked to Life Cycle Assessment (LCA) applications. LCA is considered as an accounting model and a social planner's view on environmental issues, rather than the minimization of a company's direct environmental liabilities. The LCM framework represents the Triple Bottom Line (3BL), that is a model that integrates the "three dimensions of sustainability: economic, environmental and social" and covers the three Ps: people, planet and profit.

Wang and Huang (2021) conducted a bibliometrics analysis of the impact of COVID-19 on sustainability and SDGs. The results revealed COVID-19 pandemic had negative effects on 17 SDGs goals, whereas the pandemic may also bring opportunities to another 14 SDGs goals. The effects of the coronavirus are not limited to SDG1 (no poverty) and SDG2 (zero hunger). Like any other disease, this virus negatively inflicts health systems and threatens SDG3 (good health and well-being). This can be attributed to the fact that hospitals and other health facilities are overburdened in many countries.

Some research studies focused on studying the impact of the coronavirus disease on SDG 3. For instance, Filho et al. (2020) discussed the impact of the coronavirus epidemic on SDGs implementation worldwide. The findings reported the priority of COVID-19 for many health systems in developing countries. However, this might disrupt preventive initiatives for other diseases and overlook mental health problems. As a result, COVID-19 might endanger the process of implementing the SDGs. This research is important to track the SDGs implementation so that the attained progress is not jeopardized.

Several research studies addressed the COVID-19 impact on multiple SDGs within the 2030 Agenda for Sustainable Development. For example, Nundy et al. (2021) investigated the global impact of the pandemic on the economic, energy, environment, transport, and human life sectors. These industries had been affected by the lockdown and social distancing strategies forced to curb the virus transmission. Most business operations ceased during the lockdown, significantly reducing energy consumption and improving the environment. Besides, the stagnant economy interrupted the human mindset and caused financial setbacks. The transportation sector attempted to recover from unsustainable losses without government assistance and friendly strategies.

Furthermore, priority shall be given to maintaining the quality of the indoor environment since many people have shifted to work remotely. The energy-efficient buildings also played an important role in reducing the building energy demand. The authors finally recommended collaborative assistance among all nations to meet the SDGs by 2030.

Ranjbari et al. (2021) examined the COVID-19 effect on the triple sustainability pillars (namely, economic, social, and environmental perspectives) and the SDGs. The study identified the current research gaps and proposed some research directions for sustainable development: 1) an action plan that modifies sustainability objectives and targets and creates a measurement mechanism based on the virus implications, 2) creative approaches for economic resilience, with a focus on SDGs 1 and 8, and 3) quantitative research expansion to harmonize the COVID-19-related sustainability research. In another study, Ranjbari et al. (2021) mapped the SDG objectives and Iran's implementation of the 2030 Agenda for Sustainable Development on a fuzzy action priority surface using a mixed-method methodology. As a result, the recovery Agenda for Sustainable Development included the following actions: cutting in half the proportion of the poor (SDG 1.2), development-oriented policies to support creativity and job creation (SDG 8.3), and stopping pandemics and other epidemics (SDG 3.3).

Shulla et al. (2021) examined the impact of the COVID-19 outbreak on the interdependencies across SDGs based on focus group discussions. The research was limited to studying five SDGs, including health and wellness (SDG3), economic growth (SDG8), and climate action (SDG13). The findings indicated the interconnection between SDGs and COVID-19 outcomes. Ameli et al. (2023) attempted to provide a policy response to achieve the SDGs while considering the long-term effects of COVID-19. The findings revealed that using any individual or combined strategy would lessen the COVID-19 impact on the SDGs in case of a medium pandemic activation level. In addition, employing a single strategy with a high activation level better achieved the SDGs than using a mix of strategies during low or medium pandemic activation levels.

8.3 SUSTAINABLE CONSUMPTION INDEX

With the mission to provide a greater understanding of the influences of the COVID-19 pandemic on sustainable consumption, an international online survey undertook with consumers to identify how the COVID-19 pandemic might have changed their consumption habits and reveal possible trends in the future. To achieve the maximum number of possible answers, a combination of convenience and snowball samples was chosen, being a non-probabilistic approach. Despite the limitations of this method, namely, the risk of bias and the lack of generalization, convenience samples are cheap, efficient, and simple to implement. When combined with a snowball approach, the sample can reach different populations and social groups, providing a diversity of viewpoints in a very small amount of time. Furthermore, when facing an emerging topic, such as COVID-19, this sample allows one to obtain the results faster and to provide up-to-date evidence.

The survey instrument was designed based on previous literature that discusses consumption patterns and attitudes and their association with the COVID-19 pandemic. Zwanka and Buff (2021) designed a conceptual framework to review the potential influence of the COVID-19 worldwide pandemic on consumer traits, buying patterns, psychographic behaviors, and other marketing activities. O'Meara et al. (2022) analyzed the consumer experience of food environments and food acquisition practices during COVID-19. The hypothesis that guided the survey development was that the COVID-19 pandemic has influenced sustainable consumption. The study, therefore, looked at three main aspects: levels of consumption during the pandemic, changes in patterns due to the pandemic; and measures being implemented to make consumption more sustainable.

Besides the descriptive statistic, used mostly on the characterization of the sample, the results were analyzed following two main approaches. Principal component analysis was used to cluster the items of subscales into principal components, in which the items clustered are highly correlated with each other.

The results of the principal component analysis were used to create four indexes:

1. Sustainable consumption induced by COVID-19 pandemic (SCI-Covid19)
2. Ecological awareness
3. Habitual Pro-Environmental Behavior
4. Occasional Pro-Environmental Behavior.

The questions in each index had a five-point Likert scale as response options. Then, individuals who answered "strongly agree" to each question obtained a maximum score (5 points multiplied by the number of questions), conversely, individuals who answered "strongly disagree" to all questions received the minimum score (1 point multiplied by the number of questions);

As the indexes are measured at different scales, we standardized them so that their values range from 0 to 1:

$$\text{Index } Ji = s_i - s_{min} / s_{max} - s_{min}, \tag{1}$$

where s_i is the original score for individual i for each variable in the index J; and s_{min} and s_{max} are the minimum and maximum score values, respectively, for each variable in the index J among all the individuals.

Then, an ordered logit model was estimated to analyze the main variables that affect the probability of pro-environmental consumption behavior as a consequence of the pandemic COVID-19. The ordered logit model is adequate because the pro-environmental consumption is based on a qualitative description. The dependent variable was based on the first component of the principal component analysis, that is, the SCI-Covid19 Index. The dependent variable comprises three levels of sustainable consumption behavior: low (if SCI-Covid19 < 0.5), medium (if $0.5 \leq$ SCI-Covid19 < 0.75), and high (if SCI-Covid19 $\geq$ 0.75). The independent variables were composed of sociodemographic features (age, gender, education, and income) and proxies for ecological awareness and two different categories of pro-environmental behavior as defined by Lavelle et al. (2015) and obtained by the principal component analysis (Ecological Awareness, Habitual and Occasional Pro-Environmental Behavior). The econometric model description can be found in Vicente-Molina et al. (2018).

Based on econometric strategies, this can be understood as a step forward, as it intends to model the pro-environmental behavior adopted by society during the pandemic onto a "Sustainable consumption induced by the COVID-19 pandemic" Index (SCI-Covid19). To this end, in this phase, two econometric multivariate statistical analysis techniques are used – the principal component analysis and the ordered logit model. The principal component analysis are aimed at grouping 23 questionnaire items into four latent variables. Subsequently, the analysis employs an ordered logit model to draw inferences about the model designed.

The data gathered has shown some interesting trends. First, the increased consumption triggered by the pandemic has been paralleled by a noticeable shift towards sustainable consumption. It is interesting to note that the smallest group is the group that shows a high sustainable consumption index, followed by the group with a medium index. Therefore, even though the pandemic alone cannot be regarded as a "game changer", it seems that it has offered an opportunity to convince those within the medium index.

A further trend identified by the study is related to the fact that some barriers seem to prevent the respondents from engaging in sustainable consumption. Some of the reasons given – namely, a lack of trust about the true sustainability of some products, the problems seen in financially affording some sustainable products, and the difficulty in finding sustainable products in the cities they live – indicate that even when willing to engage in more sustainable consumption patterns, consumers were deterred by these problems.

8.4   LIFECYCLE SUSTAINABILITY ASSESSMENT (LCSA)

LCSA is an evaluation procedure that measures product performance with regard to environment, economy, and society. The accepted model for applying LCSA is defined by Equation (2) (Traverso et al., 2012):

$$LCSA = LCA + LCC + S\text{-}LCA \qquad (2)$$

The formal equation indicates that to assess the sustainability performance of products over the whole life cycle, a complementary approach of environmental Life Cycle Assessment (LCA), Life Cycle Costing (LCC), and Social Life Cycle Assessment (S-LCA) should be implemented with a consistent system boundary and by reporting all results in the same functional unit. Whereas LCC assesses economic impacts (Swarr et al., 2011), S-LCA is the assessment of the positive and negative social influences of products.

The term resilience has been used freely across a wide range of academic disciplines and in many different contexts. There is little consensus regarding what resilience is, what it means for organizations and, more importantly, how organizations might achieve greater resilience in the face of increasing threats. The aim of this method is to propose a general framework to connect resilience and sustainability assessments.

How resistant are (we and) our environment to crises like COVID-19?
The incidence of the pandemic on the environment produced and is producing contrasting impacts. In response to the pandemic, partial and total lockdowns were enforced in many parts of the world. Most of the papers reviewed focus on impacts related to the environment and on air pollution and climate change (Ciciotti, 2020). These lockdowns provided an unprecedented opportunity to test how major transportation policy (Tuffs, 2020) interventions and reforms in production patterns might contribute to enhance urban air quality (Kerimray et al., 2020). In the end, it should be reiterated that this crisis highlights the need for critical reflections on the importance of cities, business and people.

What is the significance of resilience in sustainability?
In the face of ever-increasing global complexity and volatility, it is essential to move beyond a simplistic "steady state" model of sustainability. There is a direct correlation between resilience and sustainability, because major studies focus their attention on the ecosystem condition and on the resilience as a quantitative approach for its equilibrium. The bibliometric study of Barbosa (2021) has shown how risk and sustainable supply chain management is explored in the literature and identifies resilience as a new research topic. In supply chain management, it is crucial to develop and implement a resilient approach to manage the different risks of disruption. For instance, in the agri-food supply chain, risks are related to meteorological, environmental, logistic/infrastructure, market, regulatory, financial and operational/managerial factors (Zhao et al., 2011). Implementing resilience could generate competitive advantage (Bottani et al., 2019).

Sustainable production represents a strategy that holistically considers economic, environmental and social impacts (Folke et al., 2010). Indeed, a company with sustainable production is resilient by respecting the environment and promoting global management. Through international standards, it is possible to implement more environmentally friendly production within companies.

Is there a link between the quantification of resilience and Life Cycle Assessment?

No, there is no evidence of the Life Cycle Assessment approach in the quantification of resilience. The evolution of the studies conducted at this time reflects different aspects of the pandemic crisis but there is research with an overview of the different aspects or that include a global perspective and analysis of the matter. The bibliometric analysis underscores this consideration, and the ATA also confirmed the bibliometric analysis results included. Now, relevant themes, such as the direct relationship between resilience and sustainability are not taking the view that the Covid-19 sustainable production can generate greater resilience favorable to both workers and the environment (for example, by reducing the impact of outsourcing) (Diaz-Elsayed et al., 2020). To be resilient, a company introduces ideological and operational changes focusing on sustainability and favoring the continuity of operations. Diaz-Elsayed et al. (2020) proposed the formalization of Continuity Of Operations Programs (COOP) such as those promoted within the Department of Defense and the Federal Emergency Management Agency (FEMA, 2020). Like the ISO 9001 certification, these certificates could encourage manufacturers to adopt sustainable, resilient practices. The COOP certificates identify companies with skills and equipment within their supply chains capable of managing and reducing risks (Diaz-Elsayed et al., 2020).

Furthermore, the increase in waste emerges among the consequences generated by the health crisis. In light of this, implementing materials and components' recovery, recycling, and reuse processes support long-term resilience and sustainability (Diaz-Elsayed et al., 2020). In addition, collaborative technologies such as blockchain technology support building resilience through sharing information (Saberi et al., 2019).

8.5 SUSTAINABILITY RANKING TECHNIQUES

The proposed flowchart for assessing the SDG indicators in reflecting COVID-19 impact is shown in Figure 8.1. The framework starts with reviewing the SDGs within the research scope and identifying the indicators for each assessed SDG. Several experts in the domain have been selected and contacted to gather their responses to assess the degree of importance of the SDG indicators. Decision matrices are then formulated for each SDG based on the perceptions of different experts. Based on the gathered responses, three ranking techniques (namely, Relative Importance Index (RII), Weighted aggregated sum product assessment-Technique (WT), and Fuzzy Analytic Hierarchy Process (FAHP)) are employed to provide initial rankings for the indicators. Various procedures are employed in different decision-making methods, resulting in distinct rankings. As a result, it is important to employ an ensemble approach in the final stage to provide an aggregated ranking for the indicators and prioritize their influence in reflecting the COVID-19 impact.

Questionnaire surveys were undertaken with key experts to assess the COVID-19 impact on indicators of specific SDGs (namely, SDGs 1, 2, 3, 8, and 13) in Africa. The

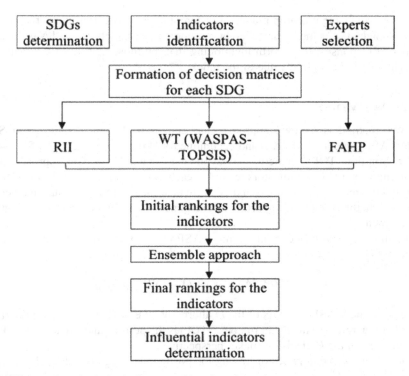

FIGURE 8.1   Sustainability indicator ranking flowchart.

questionnaire was divided into four sections: an informed consent form, respondent information, questions, and indicator descriptions. The respondents were asked to assign only one score that ranged from "1-lowest importance" to "9-highest importance" for each indicator in the five SDGs. These scores represent the weights of importance of the assessed indicators from the perspective of experts. Besides, they were encouraged to assign distinct scores to different indicators. The online delivery method using Google Forms was chosen to maintain social distancing and eliminate negative consequences. The main stakeholders were selected based on their roles and responsibilities in implementing SDGs.

This section provides background on the four applied ranking techniques of indicators; relative importance index, WT, fuzzy analytic hierarchy process, and ensemble ranking techniques.

8.5.1   RELATIVE IMPORTANCE INDEX

The RII technique was developed by Aibinu and Jagboro (2002) to describe the relative significance of different alternatives, as per Equation (3).

$$RII = \Sigma w / (A \times N) \qquad (3)$$

where; w is the weight of importance of different alternatives from the perception of each respondent, A is the highest weight of importance (namely, 9), and N is the total number of responses. It is worth mentioning that the most significant alternative is associated with the highest value of the index.

8.5.2 WT Method

The WT method is a novel approach that combines Weighted Aggregated Sum Product ASsessment (WASPAS) and Technique for Order of Preference by Similarity to Ideal Solution (TOPSIS) methods (Davoudabadi et al., 2019). This approach ranks the alternatives (namely, indicators related to each SDG) using weights of importance of attributes (namely, responses) and performance measurements of alternatives. In this research, the weights of importance of responses are measured using the Shannon entropy method.

The first step involves computing the WASPAS index for each alternative, as per Equation (4).

$$\psi_i = \Gamma \Sigma^K_{k=1}(\varphi^{(k)}(A_i) \times W_j) + (1 - \Gamma)\Pi^K_{k=1}\varphi^{(k)}(A_i)^{Wj} \tag{4}$$

where; ψ_i is the WASPAS index of the ith alternative, $\Gamma \in [0,1]$, W_j is the weight of the jth attribute, A_i refers to the ith alternative, $\varphi^{(k)}(A_i)$ is the performance measure for the ith alternative from the kth decision-maker perspective.

The second step comprises of finding the positive and negative ideal alternatives, as per Equations (5), (6)).

$$r^+ = min_i(\psi_i) \tag{5}$$

$$r^- = max_i(\psi_i) \tag{6}$$

where; r^+ and r^- are the positive and negative ideal alternatives, respectively.

The last step considers computing the closeness coefficient for each alternative, as per Equation (7).

$$C_i = \frac{\left|\psi_i - r^-\right|}{\left|\psi_i - r^+\right| + \left|\psi_i - r^-\right|} \tag{7}$$

where; Ci is the closeness coefficient for each alternative. It should be noted that a superior alternative is linked to a higher closeness coefficient value.

8.5.3 Fuzzy Analytic Hierarchy Process

An improved FAHP is utilized to quickly collect and analyze responses from many experts. This method is based on a new survey that assesses the alternatives in a row instead of performing pairwise comparisons between every two alternatives. The improved questionnaire survey includes all factors in the first column and ratings in the other columns. It is beneficial for developing a consistent judgment matrix based on diverse responses on a complicated topic, including many factors.

The procedures for employing the triangular FAHP technique to rank the indicators for each SDG are summarized as follows:

1. Allocating different scores to each indicator
2. Representing each score by an interval number ranging from 1 to 9, and summarizing the lowest and highest scores attributed to this indicator
3. Establishing a judgment matrix by comparing two indicators in the same layer pairwise and computing each coefficient as a ratio of the two-interval values for the two indicators
4. Replacing the previously specified ratio with a crisp value that meets the consistency criterion of the judgment matrix
5. Replacing the crisp value with a triangular fuzzy number and creating a judgment matrix.

Finally, Chang (1996) extended analysis has been used in this research to compute the weights of importance of the indicators from fuzzy pairwise comparison matrices because of its popularity in FAHP algorithms.

8.5.4 ENSEMBLE RANKING

A novel method based on half-quadratic theory is used to calculate the ensemble ranking of decision-making techniques, as per Equations (8), (9), (10) (Mohammadi and Rezaei, 2020).

$$\alpha_m = \delta(\|R_m - R^*\|_2) \tag{8}$$

Where; α_m is the half-quadratic auxiliary variable, m is the number of decision-making techniques, R_m is the ranking of the mth decision-making technique, and R^* is the final aggregated ranking.

$$w_m = \alpha_m / \Sigma_j \alpha_j \tag{9}$$

where; w_m is the weight of the mth decision-making technique.

$$R^* = \Sigma_m w_m \times R^m \tag{10}$$

Equation (11) calculates the consensus index, which represents the degree of agreement among decision-making techniques on the final ranking.

$$C(R^*) = \frac{1}{KM} \sum_{k=1}^{K} \sum_{m=1}^{M} \frac{N_\sigma(R_k^* - R_k^m)}{N_\sigma(0)} \tag{11}$$

where; $C(R^*)$ is the consensus index of the final ranking, K is the number of alternatives, and N_σ is the probability density function of the Gaussian distribution with a standard deviation of σ and a mean of zero.

Finally, Equation (12) is used to evaluate the trust level, which determines the degree to which the ensemble ranking may be certified.

$$T(R^*) = \frac{1}{K} \sum_{k=1}^{K} \sum_{m=1}^{M} w_m \times \left(\frac{N_\sigma(R_k^* - R_k^m)}{N_\sigma(0)} \right) \tag{12}$$

where; $T(R^*)$ is the trust level of the aggregated ranking.

8.6   RESULTS AND DISCUSSION

The computation procedures used in the RII, WT, and FAHP methods for the SDG1 indicators are implemented using MATLAB R2019a and Microsoft Excel. As for the RII, the numerator is calculated by considering the respondents' frequency and weight assigned to each indicator. The rankings obtained from the RII, WT, and FAHP methods are finally aggregated using an approach based on the half-quadratic theory. The rankings are obtained from the RII, WT, and FAHP methods, as well as the ensemble rankings for SDG1 indicators. The ensemble ranking is determined by minimizing the Euclidean distance to each calculated ranking. The results indicate that the share of the population living below the international poverty line is the most influential indicator among SDG1 indicators in reflecting COVID-19 impact. The final ranking has a consensus index of 0.85, indicating a high degree of agreement among the rankings. Besides, it is associated with a trust level of 0.96, implying that the final ranking could be accredited.

The indicators of the remaining SDGs are computed similarly. It is worth noting that the ensemble rankings for SDG2, SDG3, SDG8, and SDG13 (see Table 8.1) obtain consensus indices of 0.97, 0.99, 0.98, and 0.85, respectively. This indicates the significant degree to which all decision-making techniques agree upon the final rankings. Meanwhile, the rankings are associated with trust levels of 0.97, 0.99, 0.98, and 0.95, respectively. This implies that the aggregated rankings for all the assessed SDGs are reliable and trusted.

It is found that the top-five most significant indicators in reflecting the COVID-19 impact on SDG1 are the share of population living below the international poverty line, the population having access to basic services, the population living below the national poverty line, the proportion of government expenditure on vital services, and pro-poor public social spending. Besides, this virus has the greatest influence on the following SDG2 indicators; undernourishment, food insecurity, average income of food producers, malnutrition among children under the age of 5, and food price anomalies. Moreover, COVID-19 has the most significant effects on some indicators related to SDG3, namely official health sector support, essential health services exposure, medical staff density, the fraction of health facilities with affordable and accessible medicines, and the portion of the population with high health-related expenditures. For SDG8, the most influential indicators in reflecting COVID-19 are reported as follows: annual GDP per capita growth rate, inflation rate, tourism share in GDP, unemployment rate, and the fraction of informal non-agricultural employment. Finally, the following are the most important indicators relevant to SDG13: number

TABLE 8.1

Symbols and descriptions of the indicators for each assessed SDG

SDG	Indicator symbol	Indicator description
SDG1	I1-1	Share of population living below the international poverty line
	I1-2	Share of population living below the national poverty line
	I1-3	The ratio of population covered by social protection
	I1-4	The ratio of the population having access to basic services
	I1-5	Number of disaster-related deaths
	I1-6	The portion of government expenditure on vital services
	I1-7	Grants attributed to poverty reduction
	I1-8	Disaster-related direct economic loss as a percentage of global GDP
	I1-9	Budgeting specified for the poor people
SDG2	I2-1	Malnutrition prevalence
	I2-2	The population prevalence of food insecurity
	I2-3	Stunting prevalence in children under the age of five
	I2-4	Malnutrition prevalence in children under the age of five
	I2-5	Anaemia prevalence in women aged 15–49 years
	I2-6	Production volume per labour unit
	I2-7	Small-scale food producers' average income
	I2-8	The proportion of agricultural land dedicated to profitable and sustainable farming methods
	I2-9	Agriculture orientation index for government spending
	I2-10	Export incentives for agricultural products
	I2-11	Anomalies in food prices
	I2-12	Agriculture sector government aid
SDG3	I3-1	The ratio of maternal mortality
	I3-2	Rate of neonatal mortality
	I3-3	The mortality rate among children under the age of five
	I3-4	The number of people killed in traffic accidents
	I3-5	Number of people dying as a result of contaminated water
	I3-6	Coverage of healthcare services
	I3-7	Aid to the health sector from the government
	I3-8	The density of health workers
	I3-9	Percentage of births attended by competent health professionals
	I3-10	The number of new HIV infections per 1000 healthy people
	I3-11	Incidence of tuberculosis per 100,000 people
	I3-12	Malaria prevalence per 1000 people
	I3-13	Incidence of Hepatitis B per 100,000 people
	I3-14	The number of persons who require treatment for neglected tropical illnesses
	I3-15	Deaths caused by diabetes, cancer, cardiovascular disease, or chronic respiratory disease
	I3-16	Suicide death toll
	I3-17	Treatment approaches for drug use disorders
	I3-18	Hazardous alcohol abuse

(*continued*)

TABLE 8.1 (Continued)
Symbols and descriptions of the indicators for each assessed SDG

SDG	Indicator symbol	Indicator description
	I3-19	The proportion of women of reproductive age (15–49) who get their family planning needs fulfilled using contemporary techniques
	I3-20	The rate of adolescent births
	I3-21	The proportion of the population with high household health expenses as a proportion of total household spending or income
	I3-22	Household and ambient air pollution causing a high death rate
	I3-23	Deaths caused by unintentional poisoning
	I3-24	The prevalence of current tobacco usage among people aged more than or equal to15
	I3-25	The proportion of health facilities with a core set of relevant medications on a long-term basis
	I3-26	Capacity for international health standards and health emergency preparation
SDG8	I8-1	The growth rate of annual GDP per capita
	I8-2	Informal work as a percentage of non-agricultural employment
	I8-3	Index of the stock market
	I8-4	Consumption of domestic materials
	I8-5	The rate of unemployment
	I8-6	Percentage of youth that are not enrolled in schools
	I8-7	The proportion of children who work
	I8-8	Tourism contribution to GDP
	I8-9	The number of commercial banks
	I8-10	National labour-rights compliance level
	I8-11	Average employee hourly income
	I8-12	Rate of inflation
	I8-13	Global competitiveness index
SDG13	I13-1	Number of disaster-related deaths
	I13-2	Adaption level of disaster risk reduction strategy among local government entities
	I13-3	GHG emissions related to energy, imports, and fossil fuel exports per year
	I13-4	Enhanced public knowledge of climate change and environmental concerns as a result of government efforts
	I13-5	Amounts of funds in connection to the ongoing objective of $100 billion through 2025
	I13-6	Number of small-island developing countries and least developed countries with national adaptation plans and strategies

of disaster deaths, level of disaster risk reduction adaptation strategy, annual greenhouse gas emissions, number of small island developing nations and least developed countries with nationally determined contributions, long-term strategies, national adaptation communications and plans, and public awareness about climate change and environmental concerns.

Relative Importance Index (RII), Weighted aggregated sum product assessment-Technique for order preference by similarity to ideal solution (WT), and Fuzzy Analytic Hierarchy Process (FAHP) methods were used to prioritize the indicators. Using the modified FAHP as a multi-criteria decision-making technique helps assess the alternatives in a row instead of performing pairwise comparisons between every two alternatives. As such, while using the pairwise comparison approach, AHP cannot cope with the impression and subjectivity of the expert judgment. As a result, the FAHP approach has captured the ambiguity in expert judgment while analyzing the impact of COVID-19 on SDG indicators. Finally, the rankings were combined using a method based on the half-quadratic theory.

8.7 CONCLUSIONS

The coronavirus pandemic has taken a heavy toll on the global economy, with social and environmental consequences arising as secondary concerns. COVID-19 curtails the progress in achieving some Sustainable Development Goals (SDGs), which represent some ways by which quality of life may be restored, and the myriad difficulties related to food and water shortage or bad health conditions could be addressed. Therefore, it is of utmost importance to continuously focus on the SDGs implementation so that the progress made so far is not compromised.

Since the early days of the COVID-19 crisis, the scientific community has constantly been striving to shed light on various issues such as the mechanisms driving the spread of the virus, its environmental and socio-economic impacts, and necessary recovery and adaptation plans and policies. Sustainable development cannot be achieved without the involvement of industries and consumers for more sustainable production and consumption. A life cycle thinking approach is the best approach to assessing products and encouraging this involvement. With life cycle thinking, potential impacts of a product life cycle are assessed and evaluated. Focusing the global sustainability assessment on the life cycle approach helps to provide more complete information useful for the continuous search for new balances to achieve progress in the sustainable management of complex systems. The economic-managerial cutting appears evident in the research and the results shows that this is a problem mainly related to these disciplines.

The results showed that the share of the population living below the international poverty line, undernourishment prevalence, official health sector support, annual gross domestic product per capita growth rate, and number of disaster deaths were the most influential indicators in reflecting the COVID-19 impact on SDG1, SDG2, SDG3, SDG8, and SDG13, respectively. The SDG ensemble rankings obtained consensus indices of 0.85, 0.97, 0.99, 0.98, and 0.85, respectively. This indicated a significant agreement on the final rankings by all decision-making techniques.

BIBLIOGRAPHY

Aibinu, A.A., & Jagboro, G.O. (2002). The effects of construction delays on project delivery in Nigerian construction industry. *International Journal of Project Management*, 20(8), 593–599.

Ameli, M., Esfandabadi, Z.S., Sadeghi, S., et al. (2023). COVID-19 and sustainable development goals (SDGs): Scenario analysis through fuzzy cognitive map modeling. *Gondwana Research*, 114, 138–155.

Arcese, G., Valeri, M., Poponi, S., et al. (2020). Innovative drivers for family business models in tourism. *Journal of Family Business Management*, Vol 11 No. 4, pp. 402-422.

Arcese, G., Schabel, L., Elmo, G.C., et al. (2019). Smart city in Europe: comparative analysis between Italy and Germany development. *International Journal of Environmental Policy and Decision Making*, 2(4), 330.

Barbosa, M.W. (2021). Uncovering research streams on agri-food supply chain management: A bibliometric study. *Global Food Security*, 28, 100517.

Barnes, S.J, Diaz, M., & Arnaboldi, M. (2020). Understanding panic buying during COVID-19: a text analytics approach. *Expert Systems with Application*,169, 114360.

Bellandi, M. (2020). Some notes on the impacts of Covid-19 on Italian SME productive systems. *Symphonya. Emerging Issues in Management symphonya.unicusano.it)*, 2, 63–72.

Black, I.R, & Cherrier, H. (2010). Anti-consumption as part of living a sustainable lifestyle: daily practices, contextual motivations and subjective values. *Journal of Consumer Behaviour*, 9(6), 437–453.

Bottani, E., Murino, T., Schiavo, M., et al. (2019). Resilient food supply chain design: Modelling framework and metaheuristic solution approach. *Computers & Industrial Engineering*, 135, 177–198.

Cavallo C, Sacchi G, & Carfora V. (2020). Resilience effects in food consumption behaviour at the time of Covid-19: Perspectives from Italy. *Heliyon*, 6(12), e05676.

Chang, D.Y., (1996). Applications of the extent analysis method on fuzzy AHP. *European Journal of Operational Research*, 95(3), 649–655.

Ciciotti, E. (2020). A new territorial-industrial policy after the Covid 19 crisis. *Symphonya. Emerging Issues in Management (symphonya.unicusano.it)*, 2, 25–32.

D'Adamo, I., & Rosa, P. (2020). How do you see Infrastructure? green energy to provide economic growth after COVID-19. *Sustainability*, 12(11), 4738.

Darestani, Y.M., Sanny, K., Shafieezadeh, A., et al. (2021). Life cycle resilience quantification and enhancement of power distribution systems: A risk-based approach. *Structural Safety*, 90, 102075.

Davoudabadi, R., Mousavi, S.M., & Mohagheghi, V., (2020). A new last aggregation method of multi-attributes group decision making based on concepts of TODIM, WASPAS and TOPSIS under interval-valued intuitionistic fuzzy uncertainty. *Knowledge of Information System*s, Volume 62, Issue 4, pp. 1371-1391.

Diaz-Elsayed, N., Rezaei, N., Ndiaye, A., et al. (2020). Trends in the environmental and economic sustainability of wastewater-based resource recovery: A review. *Journal of Cleaner Production*, 7, 265.

Fazlollahtabar, H. (2016). Parallel autonomous guided vehicle assembly line for a semi-continuous manufacturing system. *Assembly Automation*, 36(3), 262–273.

Fazlollahtabar, H. (2018a). Lagrangian relaxation method for optimizing delay of multiple autonomous guided vehicles. *Transportation Letters*, 10(6), 354–360.

Fazlollahtabar, H. (2018b). Scheduling of multiple autonomous guided vehicles for an assembly line using minimum cost network flow. *Journal of Optimization in Industrial Engineering*, 11(1), 185–193.

Fazlollahtabar, H. (2019a). An effective mathematical programming model for production of automatic robot path planning. *The Open Transportation Journal*, 13(1), 11–16.

Fazlollahtabar, H. (2019b). Triple state reliability measurement for a complex autonomous robot system based on extended triangular distribution. *Measurement*, 139, 122–126.

Fazlollahtabar, H. (2020). Comparative simulation study for configuring turning point in multiple robot path planning: Robust data envelopment analysis. Robotica, 38(5), 925–939.

Fazlollahtabar, H. (2021). Robotic Manufacturing Systems Using Internet of Things: New Era of Facing Pandemics. *Automation, Robotics & Communications for Industry*, 4.0, 82.

Fazlollahtabar, H. (2022). Internet of Things-based SCADA system for configuring/reconfiguring an autonomous assembly process. *Robotica*, 40(3), 672–689.

Fazlollahtabar, H., & Hassanli, S. (2018). Hybrid cost and time path planning for multiple autonomous guided vehicles. *Applied Intelligence*, 48, 482–498.

Fazlollahtabar, H., & Jalali, S.G. (2013). Adapted Markovian model to control reliability assessment in multiple AGV. *Scientia Iranica*, 20(6), 2224–2237.

Fazlollahtabar, H., Mahdavi-Amiri, N., & Muhammadzadeh, A. (2015). A genetic optimization algorithm for nonlinear stochastic programs in an automated manufacturing system. *Journal of Intelligent & Fuzzy Systems*, 28(3), 1461–1475.

Fazlollahtabar, H., & Niaki, S.T.A. (2017a). Binary state reliability computation for a complex system based on extended Bernoulli trials: Multiple autonomous robots. *Quality and Reliability Engineering International*, 33(8), 1709–1718.

Fazlollahtabar, H., & Niaki, S.T.A. (2017b). Integration of fault tree analysis, reliability block diagram and hazard decision tree for industrial robot reliability evaluation. *Industrial Robot: An International Journal*, 44(6), 754–764.

Fazlollahtabar, H., & Niaki, S.T.A. (2017c). *Reliability Models of Complex Systems for Robots and Automation*. CRC Press.

Fazlollahtabar, H., & Niaki, S.T.A. (2018a). Cold standby renewal process integrated with environmental factor effects for reliability evaluation of multiple autonomous robot system. *International Journal of Quality & Reliability Management*, 35(10), 2450–2464.

Fazlollahtabar, H., & Niaki, S.T.A. (2018b). Modified branching process for the reliability analysis of complex systems: Multiple-robot systems. *Communications in Statistics-Theory and Methods*, 47(7), 1641–1652.

Fazlollahtabar, H., & Saidi-Mehrabad, M. (2015a). *Autonomous Guided Vehicles: Methods and Models for Optimal Path Planning*. Germany: Springer International Publishing.

Fazlollahtabar, H., & Saidi-Mehrabad, M. (2015b). Risk assessment for multiple automated guided vehicle manufacturing network. *Robotics and Autonomous Systems*, 74, 175–183.

Fazlollahtabar, H., & Saidi-Mehrabad, M. (2019). *Cost Engineering and Pricing in Autonomous Manufacturing Systems*. Emerald Publishing Limited.

Fazlollahtabar, H., Saidi-Mehrabad, M., & Balakrishnan, J. (2015a). Mathematical optimization for earliness/tardiness minimization in a multiple automated guided vehicle manufacturing system via integrated heuristic algorithms. *Robotics and Autonomous Systems*, 72, 131–138.

Fazlollahtabar, H., Saidi-Mehrabad, M., & Balakrishnan, J. (2015b). Integrated Markov-neural reliability computation method: A case for multiple automated guided vehicle system. *Reliability Engineering & System Safety*, 135, 34–44.

Fazlollahtabar, H., Saidi-Mehrabad, M., & Masehian, E. (2015). Mathematical model for deadlock resolution in multiple AGV scheduling and routing network: A case study. *Industrial Robot: An International Journal*, 42(3), 252–263.

Fazlollahtabar, H., Saidi-Mehrabad, M., & Masehian, E. (2021). Robotic industrial automation simulation-optimization for resolving conflict and deadlock. *Assembly Automation*, 41(4), 477–485.

Fazlollahtabar, H., & Shafieian, S.H. (2014). An optimal path in an AGV-based manufacturing system with intelligent agents. *Journal for Manufacturing Science and Production*, 14(2), 87–102.

FEMA, "Continuity Guidance Circular–February 2018," (2018). FEMA National Continuity Programs. available at: www.fema.gov/sites/default/files/2020-07/Continuity-Guidance- Circular_031218.pdf.

Fiksel, J. (2003). Designing resilient, sustainable Systems. *Environmental Science & Technology*, 37(23), 5330–5339.

Filho, W., Brandli, L.L., Lange Salvia, A., et al. (2020). COVID-19 and the UN sustainable development goals: Threat to solidarity or an opportunity? *Sustainability*, 12(13), 5343.

Finkbeiner, M., Schau, E.M., Lehmann, A. et al. (2010). Towards life cycle sustainability assessment. *Sustainability*, 2, 3309–3322.

Folke, C., Carpenter, S.R., Walker, B., et al. (2010). Resilience thinking: Integrating resilience, adaptability and transformability. *Ecology and Society*, 15(4), 20-31.

Goddard, J. (2021). Covid-19. civic universities, societal innovation and the recovery of local communities. *Symphonya. Emerging Issues in Management (symphonya.unicusano.it)*, 1, 56–63.

Hunt, A., & Watkiss, P. (2011). Climate change impacts and adaptation in cities: A review of the literature. *Climatic Change*, 104(1), 13–49.

Johansson, A., Kisch, P., & Mirata, M. (2005). Distributed economies-a new engine for innovation. *Journal of Cleaner Production*, 13(10–11), 971–979.

Kerimray, A., Baimatova, N., Ibragimova, O.P., et al. (2020). Assessing air quality changes in large cities during COVID-19 lockdowns: The impacts of traffic-free urban conditions in Almaty, Kazakhstan. *Science of the Total Environment*, 139179. https://doi.org/10.1016/j.scitotenv.2020.139179

Korhonen, J., & Seager, T.P. (2008). Beyond Eco-Efficiency: A Resilience Perspective. *Strategic Sustainability Management*, 17(7), 411–419.

Kou, N., & F. Zhao. (2011). Effect of multiple-feedstock strategy on the economic and environmental performance of thermochemical ethanol production under extreme weather conditions. *Biomass and Bioenergy*, 35(1), 608–616.

Lavelle, M.J, Rau, H, & Fahy, F. (2015). Different shades of green? Unpacking habitual and occasional pro-environmental behaviour. *Global Environmental Change*, 35, 368–378.

Marchand A, & Walker S. (2008). Product development and responsible consumption: Designing alternatives for sustainable lifestyles. *Journal of Cleaner Production*, 16(11), 1163–1169.

Marzouk, M., Elshaboury, N., Azab, S., et al. (2022). Assessment of COVID-19 impact on sustainable development goals indicators in Egypt using fuzzy analytic hierarchy process. *International Journal of Disaster Risk Reduction*, 82, 103319.

Mirata, M., Nilsson, H., & Kuisma, J. (2005). Production systems aligned with distributed economies: Examples from energy and biomass sectors. *Journal of Cleaner Production*, 13(10–11), 981–991.

Mohammadi, M., & Rezaei, J. (2020). Ensemble ranking: Aggregation of rankings produced by different multi-criteria decision-making methods. *Omega*, 96, 102254.

Mont, O., & Plepys, A. (2008). Sustainable consumption progress: Should we be proud or alarmed? *Journal of Cleaner Production*, 16(4), 531–537. doi: 10.1016/j.jclepro.2007.01.009

Mu, D., Seager, T.P., Rao, P.S. et al. (2011). A resilience perspective on biofuel production. *Integrated Environmental Assessment and Management*, 7(3), 348–359.

Nhamo, L., & Ndlela, B. (2021). Nexus planning as a pathway towards sustainable environmental and human health post-Covid-19. *Environmental Research*, 192, 110376.

Nundy, S., Ghosh, A., Mesloub, A., et al (2021). Impact of COVID-19 pandemic on socioeconomic, energy-environment and transport sector globally and sustainable development goal (SDG). *Journal of Cleaner Production*, Article 127705.

O'Connell, S., & Hou, D. (2015). Resilience: A New Consideration for Environmental Remediation in an Era of Climate Change. *Remediation Journal*, 26(1), 57–67.

O'Meara, L, Turner, C., Coitinho, D.C., et al. (2022). Consumer experiences of food environments during the Covid-19 pandemic: global insights from a rapid online survey of individuals from 119 countries. *Global Food Security*, 32, 100594.

Paço, A., & Laurett, R. (2019). Environmental behaviour and sustainable development. In: Leal Filho W, editor. *Encyclopedia of Sustainability in Higher Education*. Cham: Springer.

Pizzol, M. (2015). Life cycle assessment and the resilience of product systems. *Journal of Industrial Ecology*, 19(2), 296–306.

Pizzol, M., Scotti, M., & Thomsen, M. (2013). Network analysis as a tool for assessing environmental sustainability: Applying the ecosystem perspective to a Danish water management system. *Journal of Environmental Management*, 118, 21–31.

Ranjbari, M., Esfandabadi, Z.S., Zanetti, M.C., et al (2021). Three pillars of sustainability in the wake of COVID-19: A systematic review and future research agenda for sustainable development. *Journal of Cleaner Production*, 297, 126660.

Ranjbari, M., Shams Esfandabadi, Z., Scagnelli, S.D., et al (2021). Recovery agenda for sustainable development post COVID-19 at the country level: Developing a fuzzy action priority surface. *Environment Development and Sustainability*, 23(11), 16646–16673.

Saberi, S., Kouhizadeh, M., Sarkis, J., et al. (2019). Blockchain technology and its relationships to sustainable supply chain management. *International Journal of Production Research*, 57(7), 2117–2135.

Sala, S., & Castellani, V. (2019). The consumer footprint: Monitoring sustainable development goal 12 with process-based life cycle assessment. *Journal of Cleaner Production*, 240, 118050.

Sarkis, J. (2020). Supply chain sustainability: Learning from the COVID-19 Pandemic. *International Journal of Operations & Production Management*, 1(1), 63–73.

Scotti, M., & B. Vedres. (2012). Supply Security in the European Natural Gas Pipeline Network. In *Networks in Social Policy Problems*, edited by B. Vedres and M. Scotti. Cambridge, UK: Cambridge University Press.

Seager, T.P. (2008). The Sustainability Spectrum and the Sciences of Sustainability. *Business Strategy and the Environment*, 17(7), 444–453.

Severo, E.A, De Guimarães, J.C.F, & Dellarmelin ML. (2020). Impact of the COVID-19 pandemic on environmental awareness, sustainable consumption and social responsibility: Evidence from generations in Brazil and Portugal. *Journal of Cleaner Production*, 286, 1st March 2021, 124947. https://doi.org/10.1016/j.jclepro.2020.124947

Sheth, J. (2020). Impact of Covid-19 on consumer behavior: Will the old habits return or die? *Journal of Business Research*, 117, 280–283.

Shojaeifar, A., Fazlollahtabar, H., & Mahdavi, I. (2016). Decomposition versus minimal path and cuts methods for reliability evaluation of an advanced robotic production system. *Journal of Automation Mobile Robotics and Intelligent Systems*, 10(3), 52–57.

Shulla, K., Voigt, B.F., Cibian, S., et al. (2021). Effects of COVID-19 on the sustainable development goals (SDGs). *Dis Sustain*, 2(1), 1–19.

Simon, J.A. (2015). Editor's perspective–the effects of climate change adaptation planning on remediation programs. *Remediation*, 25(3), 1–7.

Swarr, T.E., Hunkeler, D., Klöpffer, W., Pesonen, H.L., Ciroth, A., Brent, A.C., & Pagan, R. (2011). Environmental Life-Cycle Costing: a Code of Practice.

Travassos, G.M, Da Cunha, D.A, & Coelho, A.B. (2020). The environmental impact of Brazilian adults' diet. *Journal of Cleaner Production*, 272, 122622. doi: 10.1016/j.jclepro.2020.122622

Traverso, M., Finkbeiner, M., Jørgensen, A., et al. (2012). Life cycle sustainability dashboard. *Journal of Industrial Ecology*, 16(5), 680–688.

Tuffs, R., Larosse, J., & Corpakis, D. (2020). Post-Covid-19 recovery policies: Place-based and sustainable strategies. *Symphonya. Emerging Issues in Management (symphonya. unicusano.it)*, 2, 55–62.

UNDP (2016) Sustainable development goals. www.undp.org/content/dam/undp/library/corporate/brochure/SDGs_Booklet_Web_En.pdf

United Nations—Department of Economic and Social Affairs. (2020). The 17 goals. https://sdgs.un.org/goals

Vicente-Molina M.A., Fernández-Sainz, A., & Izagirre-Olaizola J., (2018). Does gender make a difference in pro-environmental behaviour? The case of the Basque country university students. *Journal of Cleaner Production*, 176, 89–98.

Walker, B., Holling, C.S., Carpenter, S.R., et al. (2004). Resilience, adaptability and transformability in social-ecological systems. *Ecology and Society*, 9(2). 1-9..

Wang, Q., & Huang, R. (2021). The impact of COVID-19 pandemic on sustainable development goals–a survey. *Environmental Research*, 202, 111637.

Worthy, R., Abkowitz, M.D., & Clarke, J.H. (2015). A Systematic Approach to the Evaluation of RCRA Disposal Facilities under Future Climate-Induced Events. *Remediation*, 25(2), 71–81.

Zhang, Y., Liu, H. Li, Y., et al. (2012). Ecological network analysis of China's societal metabolism. *Journal of Environmental Management*, 93(1), 254–263.

Zwanka R.J., &Buff, C. (2021). COVID-19 generation: A conceptual framework of the consumer behavioral shifts to be caused by the COVID-19 pandemic. *Journal of International Consumer Marketing*, 33(1), 58–67.

9 IoT-Based Sustainable Production Value

9.1 INTRODUCTION

Sustainability has captured a lot of attention over the last few years. The increasing pressures from different stockholders, government regulatory activity, growth of nongovernmental organizations and movements suggest that the public is no longer satisfied with companies that solely focus on maximizing profit (Figge and Hahn, 2004). In fact, people want companies to consider holistic human needs. Therefore, a growing number of companies are considering that a sustainable strategy is necessary to acquire competitive advantages. The definition of sustainability often refers to meeting the needs of current population as well as future generations. Bernal et al. (2018) argue for the inclusion of all economic, social and environmental performance indicators to be considered in order to reach global sustainability. Thus, reducing the adverse environmental and social impact of industrial activities is no longer a luxury, but an all-important necessity. Lean is a corporate management strategy that focuses on waste reduction, improving efficiency, reducing operational costs and acquiring sustained improvement in the processes of an organization. Six Sigma is a data driven method based on an increase in performance and a decrease in process variation leading to a reduction of defects and enhanced profit. The fundamental objective of the Lean Six Sigma is to provide a statistical-based strategy that focuses on continuous improvement methodology for variation reduction in a product, process or service. Lean Six Sigma is a combination of the two powerful performance improvement methodologies. This integration is intended to overcome the defects of both. Bhuiyan and Baghel (2005) claimed that Lean Six Sigma (LSS) addressed important issues that are overlooked in separate lean manufacturing and Six Sigma individually and the fusion of the two helps organizations maximize their potentials for improvement. Lean accelerates Six Sigma when lean tools help to decrease or eliminate wastes; Six Sigma relies on variation–reduction in process. Therefore, the principles of LSS lead to continuous and faster improvement process.

Researchers have made continuous efforts to integrate the concept of sustainability into many methodologies recently in order to build a single operational model. According to Galdino De Freitas et al. (2017) the integrated Lean, Lean Six Sigma and sustainability provides a systematic alignment of approaches to handle an innovative area of study. Another reason is the need for a more practical model

DOI: 10.1201/9781003400585-9

for managing and controlling sustainability in organizations (Bastas and Liyanage, 2018. Lean, sustainability and Lean Six Sigma are three methods that are complementary and in sync with each other while each strategy has the capability to ease the shortcomings of the others. The lean paradigm is constituted for waste reduction (Hartini and Ciptomulyonob. 2015), however, it does not solely take into account the environmental and social impacts (EPA, 2006; Pampanelli et al., 2014). To fill the gap, organizations have proposed sustainability (Sharrard et al., 2008). Later, to achieve a continuous peak sustainability process, the Lean Six Sigma approach has integrated these approaches to address this gap (Banawi and Bilec, 2014; Han et al., 2008; GarzaReyes, 2015; Cherrafi et al. 2016).

As mentioned, companies integrate the concept of sustainability to achieve long term benefits. Measuring corporate sustainability is defined as the measurement of conditions in which companies incorporate three pillars of sustainability (economic, social and environment) into their activities and ultimately, measuring the effect of their activities on sustainability (Artiach et al. 2010). In fact, companies measure corporate sustainability to assess integrated sustainability performance. Considering a sustainable measurement system, indicators and bounces can help organizations make more deliberate, thoughtful decisions. Any companies aiming to be sustainable must develop a performance measurement system to ensure that measurement is aligned to sustainable strategy, and a system is working effectively in driving sustainable performance. Such a system may be useful to provide pertinent information for decision makers before they make their decisions, to promote organizational learning and to encourage stakeholder engagement (Veleva and Ellenbecker, 2001).

The main idea in developing such integration between sustainability and LSS is to configure an assessment method for the performance management purpose of manufacturing systems in the Industry 4.0 context. Based on the latest innovative Industry 4.0 technology all process data in the different process steps of a production system can be integrated in one system. Knowhow generation based on the correlation of several variables can be automatically provided including the specific weight factors and applied in a continuous improvement cycle. Thus, we aim to develop a novel assessment method for production sustainability based on lean and Six Sigma effective indices in manufacturing systems. First, a collection of lean and Six Sigma indices are investigated and categorized in a brain storming process. Then the relationships of the categories are mapped with the principal pillars of sustainability namely economic, environmental and social. Having the indices classified under the main factors of sustainability, a comprehensive assessment formula is proposed including mathematical terms to compute the overall Lean-Six Sigma-sustainability of manufacturing systems. Therefore, the major contributions of the paper are listed below:

- Collecting Lean Six Sigma indices from the literature
- Categorizing effective indices of Lean Six Sigma in Industry 4.0 via Delphi method
- Mapping factors of sustainability with categorized indices of Lean Six Sigma
- Proposing a comprehensive assessment model for integrated Lean Six Sigma sustainability in an Industry 4.0 context

9.2   RELATED WORKS

In this section we review related research produced in the past. Here, we categorize the research into separate topics to emphasize the current applications and development trends of the literature.

According to the United Nations Environment Program, sustainable production (and consumption) can be viewed as the creation (and use) of products to address human needs and bring better quality of life while the use of natural resources, toxic substances, emissions and pollutants over the lifecycle of a product are minimized in such a way that the needs of future generations are not jeopardized. Thus, the need for manufacturing industry, in addition to economic efficiency, is to assess the environmental and social objectives in advancing manufacturing operations, technologies, and the competitive position (Rosen and Kishawy, 2012). The manufacturing industry consumes a lot of limited energy, materials and services with a low energy efficiency and resource conversion rate (Cai et al. 2018), which results in a lot of waste and causes serious damage to the environment. Therefore, promoting energy, resources and service efficiency and improving environmental performance as much as possible are a major problem to be solved (Mikulčić et al. 2013), contributing to realizing the sustainable development of the manufacturing industry (Lv et al. 2018). The rapid realization of the transformation of the sustainable manufacturing paradigm has been the common choice for many countries to seize the commanding strategic position in international manufacturing competition (Cai et al. 2017). As a result, research into sustainable manufacturing has aroused extensive interest, which is becoming a hot field of interdisciplinary research (Feng et al. 2016). There is evidence of sustainable work in product design, supply chain, production technology and avoiding waste activities (Mbaye 2012), which is the only way for industrial transformation and upgrading, and important content for the construction of ecological performance.

Sustainability is based on a simple principle: Everything that we need for our survival and well-being depends, either directly or indirectly, on our natural environment. To pursue sustainability is to create and maintain the conditions under which humans and nature can exist in productive harmony to support present and future generations. Hassini et al. (2012) defined business sustainability as the ability to organize business to achieve economic long-term goals as well as environment and society. This definition implies that companies strive to satisfy various conflicting objectives, while maximizing profits, calls for reducing operational costs and minimizing environmental impact and increasing social well-being and need reduction in operational costs. Sustainability is growing significantly as a customer, societal and market need, placing sustainability management as an essential stakeholder requirement for organizations (Garvare and Johansson, 2010). Consequently, Bastas and Liyanage (2018) articulated that Sustainability Management (SM) is considered to be a fundamental element for business continuity and for the fulfillment of current requirements while not sacrificing the ability of meeting future needs.

Sustainable value integrates environmental and social dimensions into financial analyses to create high value-added products and services rather than focusing solely on manufacturing output (Tao and Yu, 2018). Sustainable value is a value-based methodology to assess sustainable performance in monetary terms. Henriques and Catarino (2015) pointed out that the sustainable value concept is using the synergies between tools from value management, value analysis, eco-efficiency and cleaner production leading to an indicator that seamlessly considers the three pillars of sustainability to create more economic value, less environmental burden, and less social burden (Figge and Hahn, 2004).

In fact, it is assumed that the lower the energy-use, the stronger the sustainable production system. To enhance the efficiency of an energy system and improve sustainable production system in a plant, a discrete event simulation method was implemented (Solding et al., 2009). Similarly, Shrirvastava and Berger (2010) have referred to sustainable production gained by addressing aspects such as resource use, energy practice, product and waste management and therefore making companies more sustainably responsive. Rosen and Kishawy (2012) considered the role of environmental sustainability in achieving sustainable manufacturing by utilizing design for environment and life cycle assessment. Rauch et al., (2016) referred to a Distributed Manufacturing (DM) concept to contribute sustainable production in emerging countries. Through the DM concept with its micro-production units, goods will be able to be delivered quickly and in a sustainable way. The Lowell Center for Sustainable Production (LCSP) defines sustainable production as producing goods and services using systems and processes that are non-polluting, conserving natural resources, economically profitable, safe for workers, healthful for societies and consumers (Alyon et al., 2017). Badurdeen and Jawahir (2017) in their research introduced the main challenges that a manufacturing company faces in order to implement sustainable manufacturing practices, and finally, some strategies that were developed to address such obstacles in terms of product, process and systems' domains to create more value to the system.

Ecolabeling supports the process of achieving sustainable production and consumption. Consumer environmental awareness helps reduce negative impacts on the environment (Wojnarowska et al., 2021). Manufacturers are facing various resource related challenges, which has slowed down the progress of Industry 4.0 adoption (Bag et al., 2021). Despite widespread recognition of the need to transition toward more sustainable production and consumption and numerous initiatives to that end, global resource extraction and corresponding socio-ecological degradation continue to grow (Mathai et al., 2021).

The lean concept provides value within process improvement and waste reduction. Lean Six Sigma, on the other hand, creates value by controlling output variability (Erdil et al., 2018). Lean manufacturing presents an integrated mechanism in an advanced manufacturing strategy to handle productivity improvement, quality maximization, and waste elimination (Resta *et al.*, 2016). Meanwhile, Six Sigma is a systematic framework for continuous improvement and new product and service development. It is mainly based on statistical approaches to control process deviation from a targeted value (Linderman et al., 2003). Galdino de Freitas et al. (2017) made use of human perceptions and concluded that Lean Six Sigma influences sustainability in organizations. Johansson and Sundin (2013) investigated the combination of lean and green product development and analyzed the similarities and differences.

Cera´volo Calia et al. (2009) studied the Six Sigma impacts on the pollution prevention program in some companies. Cluzel et al. (2010) studied the link between Six Sigma and lifecycle to provide an eco-design using a novel methodology. The proposed methodology was helpful in complex systems and projects. Also, researchers made use of Six Sigma as a continuous improvement tool in sustainable manufacturing (Zhang and Awasthi 2014). Miller et al. (2010) proposed a lean sustainability model for a furniture production industry and used discrete event simulation as well as mathematical optimization to handle all factors: financial, social and environmental. Hajmohammad et al. (2013) designed an evaluation of the effect of lean on supply chain management to study the influences of environmental practices on manufacturing organizations' performance. Banawi and Bilec (2014) proposed a framework to investigate the environmental impacts on contractors of constructions companies and integrated it with Lean Six Sigma as a continuous monitoring tool.

Industry 4.0 or the fourth industrial revolution has emerged as a hot topic of discussions for companies, researchers, as well as international governments (Gabriel and Pessel, 2016). Industry 4.0 is a digital transformation of the industry by assimilating the Internet of Things (IoT), information integration and other high-tech developments which begins with focusing on the production/manufacturing sector and expands toward many sectors beyond the industry. Industry 4.0, also known as Smart Factory 4.0, is labeled as the fourth industrial revolution, yet many don't understand it or how it will impact the things they do; Quality Management is no exception (Liao et al., 2018; Chen et al., 2018). The executives in your organization, however, are most certainly monitoring this paradigm-shifting strategy. Quality professionals should be as well. Quality professionals must ensure they are part of their company's Industry 4.0 dialogue (Putnik et al., 2015). As quality professionals forge into this new era, it is critical to have a solid understanding of the premise and aspects of Industry 4.0/Smart Factory 4.0, and its implications for production, the extended supply chain, and their quality management System. For quality management systems and professionals to excel and contribute bottom-line benefits to the organization, quality processes and data need to be an embedded as integral parts of the ecosystem. Under Industry 4.0, massive amounts of data will be available to quality personnel in real-time, and from multiple sources simultaneously. The data must be used to enable quick, situational decision-making. Quality will need to embrace data (internal and external) and technology and use them to drive innovation while improving overall quality (Mayer and Pantförder, 2014).

9.2.1 Consequences and Research Gaps

To sum up, the concept of lean and Lean Six Sigma were profoundly studied before in the production paradigm. In addition, sustainable management and sustainable production are very popular areas in research works. However, as shown in the reviewed literature, the integration between Lean Six Sigma and sustainable production was rarely investigated. Thus, one of the research directions of this paper is to propose a structure to aggregate Lean Six Sigma and sustainable production as being helpful in a tactical production strategy management. Furthermore, sustainability evaluation is considered within the concept of sustainable value in literature, but sustainable production evaluation requires the development of an appropriate framework to collect the necessary elements effective on sustainable production with respect to substantial pillars of sustainability. Meanwhile, sustainability based on Lean Six Sigma within the context of Industry 4.0 is a significant contribution of this paper. Therefore, we aim to first investigate the required elements of Lean Six Sigma, categorize them in relation to the main factors of sustainability, and formulate a sustainable production measurement method. In the past, sustainable value was developed and we aim to be pioneers in developing the sustainable production value concept in the Industry 4.0 context.

9.3   PROBLEM DESCRIPTION AND MODELING

Sustainable production is significant from several perspectives. Firstly, it helps to increase sales so that three out of four are willing to pay more for products or services from brands that are committed to positive social and environmental impact. Secondly, energy consumption can be controlled and diminished by regularly inspecting manufacturing equipment for malfunctions or poor energy use, implementing recycling initiatives, and going paperless. These are also great ways to significantly reduce energy costs. Thirdly, governmental incentives are accessible while becoming a sustainable manufacturer to increase competitive ability for government contracts, since many of these contracts are only available to sustainable manufacturers. Fourthly, sustainability also leads to increased innovation. When employees are encouraged to follow sustainable methods, they will be inspired to pursue research and development projects that create new, efficient ways of doing things. Finally, aside from the benefits that directly influence your business, this decrease in carbon footprint can make a positive difference to the safety of your employees working in these conditions and in your community.

Companies aiming to be sustainable must face the challenge of measuring sustainability by designing sustainable performance system, which provide the corporation with the information necessary to assist management in the monitoring, planning and performance of its economic, environmental and social activities.

While sustainable production is significant, we can integrate it with continuous improvement to obtain strategic visions in industrial corporations. We aim to make use of Six Sigma purified by the lean concept and then merge it into sustainability factors in production systems. After that, using mathematical formulations a measurement relation is developed.

The quality management factors are collected from the literature and purified based on leanness and Six Sigma concepts using the Delphi technique. Then, an adapted interpretive structural model is proposed to collect sustainability indices-categorize-rand and map them with Lean Six Sigma elements. The quantification for the mapped indices and elements is performed by developing a mathematical formulation for sustainable production.

First, we need to collect Lean Six Sigma elements. To do that an extensive literature survey is conducted and the outcome is inserted into a Delphi framework.

The Delphi method essentially consists of at least two rounds of data collection. In the first round, 50 experts in the quality management domain were selected randomly among different manufacturing industries to participate in the survey. To prevent group pressure and the influence of dominant people, experts anonymously answered a questionnaire. In the questionnaire, experts were asked to determine the most important elements of Lean Six Sigma. Next, all the possible factors contributing to Lean Six Sigma were compiled into a single factor list. For the second round, a questionnaire was created like the first one. The only difference was that, it contained a summary of findings from the previous round, which allowed experts on the panel to change their opinion. In the instruction of this questionnaire, the respondents were asked to rate the elements on a five-point scale, ranging from 1 the least important to 5 the most important element. The third and final round was also conducted to determine the consensus of experts. Firstly a summary of the second round was presented and then the elements that had the greatest consensus among experts were identified. In this round, the study reached a stable consensus. In other words, there weren't radical changes from the third round to the fourth round. Ultimately, the results of the Delphi method were used to predict the most significant elements of Lean Six Sigma by using the insights of an expert panel in industrial production systems.

The reason for using Delphi versus other brain storming-type methods is the consensus concentration of Delphi which is very significant in grouping the quality elements to find out the most effective one within the Lean Six Sigma context.

According to the first step of the methodology, the required Lean Six Sigma elements are collected from literature. The main Lean Six Sigma elements effective on production systems are listed below:

1. 5S
2. Kaizen
3. Lean Six Sigma (process capability index)
4. Poka-yoke
5. Training on lean concepts
6. Total productive maintenance
7. Kanban
8. Collaboration with customers
9. Alternative energy sources
10. Value stream mapping.

These elements are comprehensive enough to handle continuous improvements in production systems.

The next step is to collect sustainable production factors and the corresponding indices and map them with the Lean Six Sigma elements given above. Now, the sustainable factors' indices need to be collected and ranked. The main sustainability factors are economic, environmental, and social. Here, the aim is to collect indices in relation to the Lean Six Sigma elements. To collect and categorize sustainable production indices corresponding to each factor an Interpretive Structural Modelling (ISM) technique is adapted.

9.3.1 ADAPTED INTERPRETIVE STRUCTURAL MODELLING

In order to structure a comprehensive system among a set of different and directly or indirectly related indices, Adapted Interpretive Structural Modelling (AISM) has been used. This methodology tried to identify the relationships between Lean Six Sigma and sustainability. The steps of the AISM are given below.

In AISM a systematic and logical thinking approach can be utilized to model a complex issue.The procedure for the AISM technique is:

Step 1 Structural Self-Interaction Matrix (SSIM). Initially, to identify relationships between three pillars of sustainability and element of Lean Six Sigma, a SSIM matrix was developed for each element that was found in the last stage. In this study, experts from 50 manufacturing industries (such as dairy products, home appliances, automobile products, wooden products, the forging industry, the food industry, the telecommunication industry, and so forth.) were consulted. In each industry, 20 companies were investigated and the required data was collected. Data was analyzed based on the relationships between any two sustainability factors and Lean Six Sigma elements in industries and then correlation of the ith (row) index with respect to the jth (column) index was presented in a matrix. Finally, the relationships were denoted by the following symbols:

V: index i will help to amend index j.
A: index j will be amended by index i.
X: index i and j will help to achieve each other.
O: index I and j are unrelated.

Step 2 Reachability Matrix. In this step the SSIM has been converted to a reachability matrix by assigning 0 and 1 to the contextual relationships. It is assumed that Lean Six Sigma elements don't influence each other, therefore the numerical sign for the relations was considered to be 0 and other relationships were substituted as follows: If the (i, j) sustainability indices are V, the reachability matrix of (i, j) is 1 and (j, i) is 0. If the (i, j) entry in the SSIM matrix is A, the reachability matrix (i, j) is 0 and (j, i) is 1. If the (i, j) entry in SSIM is X then the reachability matrix (i, j) is 1 and (j, i) also is 1. Ultimately, O for both (i, j) and (j, i) entry becomes 0 in the reachability matrix.

Step 3 Level Partition. From the reachability matrix, elements which help to achieve sustainability were driven. The intersection sets for each factor of sustainability, environmental, social and economic were listed in a matrix. Elements which facilitate the environmental problem have been assigned to 1 regarding the mentioned column and categorized as a set of elements which may aid sustainability in the environmental index.

After finalizing critical success factors for the implementation of Lean Six Sigma and three pillars of sustainability, AISM has been used to find out level of the factors.

For 5S Lean Six Sigma elements, several indices are categorized in the main three sustainability factors, environmental, social and economic. The same process is repeated for all Lean Six Sigma elements and the obtained sustainability factors and the corresponding indices related to Lean Six Sigma elements are inserted for the next stage called, sustainability measurement. The proposed AISM is an adaptation of ISM with four new sub-steps of including sustainability indices, categorize, rank, map with Lean Six Sigma elements.

9.3.2 PROPOSED SUSTAINABLE PRODUCTION-QUALITY IN THE INDUSTRY 4.0 CONTEXT

The system provides a new database using powerful algorithms to get more out of the existing data. The database can store production and quality engineering and management data for a long period of time. Before storing data, it is always verified and validated by a dynamic rule-based verification set. Thereafter, the resolution is adapted to the needs of the quality and production team. As a next step, data is linked together with context and especially genealogy. Finally, the data packages are compressed without any loss and stored for all process steps of the quality system. A configuration of a production system in Industry 4.0 context is depicted in Figure 9.1.

This specific preparation and storage of data allows the user to get answers to queries in seconds about all different process steps. It is often more than one variable which has contributed to deviations in quality. Also, it turned out that frequently the contributing variables were located in different process steps as they influence the sustainability of the production system. Each variable contributes a different weight

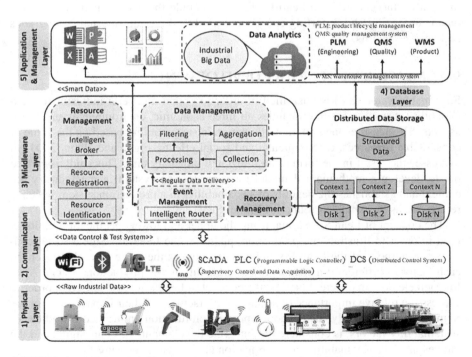

FIGURE 9.1   A Structural diagram of a production system in an Industry 4.0 context.

factor to the deviation in quality. This is an indication of the severity and also a clear direction for corrective action.

The data correlator module automatically provides all this necessary information as a transparent decision tree. The decision tree is automatically calculated using neuronal networks and indicates all weighing factors for different levels of the tree, which can be used for the setting of rule thresholds. The module even automatically provides a proposal for a new rule set which can be used in the quality decision in each process step. Customer claims are used to generate new rules or a complete new "if/than" rule set.

Once the knowhow is generated, it cannot get lost after a period, but will be permanently stored in the rule set of the system. Knowledge can be retrieved from the system easily for future "on the spot" analyses or for continuous adaptation to changed conditions. For the first time this system configuration offers continuous adaptation to the ever faster changing requirements of markets and especially high-end customers. And this is not only related to frequently changing product specifications, but also to smaller orders down to 1 product lots, which need a very flexible and short-term adaptation of process and quality control. Fast and positive acceptance by all quality system users can only be achieved if the system fulfills all ergonomically and current usability requirements. Ease of use allows each quality and production engineer to maintain and operate the system completely autonomously. This includes the generation of new and the maintenance of existing specifications, norms and customer specific requirements. Besides the quality targets, all modifications have to include the economical view, which means analyzing and reporting exactly the produced output, internal downgrades and verified claims.

9.4   MEASUREMENT MODEL DEVELOPMENT

Sustainable Value (SV) is quantified in order to obtain a general indicator for sustainability measurement that includes all three economic, environmental and social aspects. SV measures the system participation in sustainability based on the value of consumable resources, environmental and social effects (Figge and Hahn, 2004). It expresses the result in the form of an integrated monetary unit. This study is carried out to develop a mathematical model for a production system in the Industry 4.0 context and establishes the benefits of sustainability in a production system which is referred to as Sustainable Production Value (SPV).

The concept of Lean Six Sigma, using the upper and lower specification limits, is used in order to illustrate the production quality as the Lean Six Sigma social context. The purpose of the Lean Six Sigma is to express the value of each element in each factor in different dimensions of sustainability in the production system. Also, for each element in different factors, the concept of benefit to cost is used in order to express the relative importance of sustainability factors and their impact on Lean Six Sigma elements. Finally, the economic production function, influenced by indices of sustainability factors, is considered to integrate the elements of Lean Sigma and the indices of each sustainability factor. Equation (1) is used to measure SPV.

$$SPV = \sum_{k}\sum_{j}\left(\left(\frac{USL_{jk} - LSL_{jk}}{6\sigma_{jk}}\right) + \left[\sum_{t=1}^{T}\frac{\left(B_{jk} - C_{jk}\right)}{\left(1+r\right)^{t}} * \left(\prod_{j} A_{jk} L_{ik}{}^{\beta_{ik}} \kappa_{ik}\right)\right]\right). \quad (1)$$

where,

j	Element of lean-Lean Six Sigma
k	Sustainability factors
i	Indices of sustainability factors
t	Time period
USL_{jk}	Upper specification limit of the jth Lean Six Sigma element with respect to the kth sustainability factor
LSL_{jk}	Lower specification limit of the jth Lean Six Sigma element with respect to the kth sustainability factor
σ_{jk}	Standard deviation of the jth Lean Six Sigma element with respect to the kth sustainability factor
B_{jk}	Benefit of the jth Lean Six Sigma element with respect to the kth sustainability factor (Unit of money)
C_{jk}	Cost of the jth Lean Six Sigma element with respect to the kth sustainability factor (Unit of money)
r	Interest rate
A_{jk}	Productivity value with respect to the jth Lean Six Sigma element of the kth sustainability factor
L_{ik}	Workforce (human-hours) required for the production system with respect to the ith index of the kth sustainability factor
β_{ik}	Elasticity of the ith index of the kth factor based on the human workforce
κ_{ik}	Required investment of the production system with respect to the ith index of the kth sustainability factor (Unit of money)

In practice, the first term of equation (1) is computed using the collected observations of Lean Six Sigma elements with respect to each sustainability factor. The dimension is a numerical value. On the other hand, the second term is composed of two relations. The first term computes the benefit to cost analysis for implementing Lean Six Sigma in factors of sustainability. The second term computes the production productivity in the production system including the indices of sustainability factors.

9.5 IMPLEMENTATION

The proposed methodology is implemented in an industrial cluster of Mazandaran, located in Northern Iran. Fifty companies included in the industrial cluster are manufacturing systems producing different types of products. The cluster's policymakers aim to configure a comprehensive quality management paradigm in order to control all aspects of the quality attributes using a remote mechanism of Industry 4.0. The era of Industry 4.0 motivates industries to make use of the benefits provided by devices and systems. Its quality equivalent, Quality 4.0, has left indelible marks on the Life Sciences industry. However, Quality's participation in Industry 4.0 has its

TABLE 9.1
Summary of indices resulting from Delphi and AISM

No.	Lean Six Sigma	Environmental	Social	Economic
1	5S	3. A lean workshop will quickly show a leak in a system, where resources are being wasted 6. Can assist companies to improve energy and materials efficiency by reducing space required for the operation and calling attention to environmental wastes 7. Reduces the chance that materials expire or become off-specification before they can be used and then require disposal 10. Help to keep an organized and Lean workplace, which can decrease the use of natural resources	1. encourage people to correctly eliminate undesirable objects 8. Helps to improve the company's handling and storage of hazardous materials and waste 11. Can help organizations to reduce risks, improve waste management, minimize risks to the health and safety of workers and the environment by providing clean and accident-free work areas	2. Improved production rate 4. Reducing document search time 5. Reduction of traveling time 9. improving processing speed in IM
2	Kaizen	1. Provides a problem-solving culture with scientific and structured thinking, which will help organizations to resolve environmental issues 2. It helps to reduce material wastes and pollution 3. Can serve as the driving force for reducing different environmental impacts generated by processes	1. Develops the engagement of employees and unleashes their creativity leading to the promotion of innovation for environmental and social progress 2. ensures a safe and healthy place to work 3. Instilling mutual trust 4. Workforce driven collaborative environment 5. Employees suitable as team member and leader 6. Shared Knowledge and responsibility	

3	Lean Six Sigma	2. Focuses on reducing defects to improve product quality, which helps to reduce environmental waste (i.e., material, water and energy consumption and waste generation)	1. Presents an effective methodology for problem solving and decision-making. It can help managers and leaders to understand and solve the environmental and social problems 2. Helps to reduce the potential accidents, leading to safer and healthier working conditions for the operators	1. Reduces waste, scrap and improves quality
4	Poka-yoke (error proofing)	1. Contributes to minimize defects therefore reducing resources consumption (energy, water, raw material) and emissions, and so forth. 2. Identifies waste reasons and improvement Steps	1. Improves firm's quality image 2. Creates Responsible Manager	
5	Training on Lean Concepts	1. Types of wastes	1. Knowledge base on value creation 2. Knowledge base on value improvement 3. Mutual trust-based relationship 4. Knowledge on collaboration 5. Knowledge on empowerment	1. Training on quality of distribution and delivery
6	Total Productive Maintenance	1. Reduces waste, scrap and improves quality	1. Empowered Employee	1. Improved production rate 2. Increased machine life 3. Increased machine availability
7	Kanban/Pull JIT	1. Reduce the different wastes result from deteriorated, damaged and spoiled products 2. Lead to a slight increase in energy, water consumption and hazardous waste volumes	1. Provide workshop space utilization 2. Improved information quality and visibility	

(continued)

TABLE 9.1 (Continued)
Summary of indices resulting from Delphi and AISM

No.	Lean Six Sigma	Environmental	Social	Economic
		3. Facilitate identification of failures and unnecessary movements in the different production processes, which allows organization to reduce the resources consumption and wastes		
8	Collaboration with Customer		1. Workforce driven collaborative environment 2. Ensured business continuity and supply quality 3. Collaborative product design and development	3. Ensured business continuity and supply quality 4. Inputs for quality improvement
9	Alternative Energy Source	1. Lower energy consumption 2. Lower emission 3. Lower water consumption 4. Lower pollutant and harmful substances	1. Organized and safe workplace 2. Improves firm's quality image	1. Increased machine capacity 2. Improved machine life
10	Value Stream Mapping (VSM)	1. Facilitates identification and visualization of no value added in the production line, thus helping organizations to avoid excess consumption and environmental waste such as water, energy consumption, and solid and hazardous waste and air emission 2. Through VSM, operators became aware of environmental impacts of production processes. This leads to identifying the best method of using the different resources, which allows organizations to recognize important environmental benefits 3. VSM can be used as a support technique for life cycle assessment	1. VSM is one of the best visual tools that could be used to improve communication with stakeholders to understand the generation and flow of value and environmental wastes during processes. 2. VSM could help to improve ergonomics, worker health and safety 3. Flexible, Responsive 4. Empowered employee	1. improved production rate 2. Reduction of inventory 3. Cells and cell-based collaborative team

TABLE 9.2
Input data for Lean Six Sigma elements in relation with sustainability factors

Lean Six Sigma element	Sustainability factors														
	Environmental					Social					Economic				
	USL_{jk}	LSL_{jk}	B_{jk}	C_{jk}	A_{jk}	USL_{jk}	LSL_{jk}	B_{jk}	C_{jk}	A_{jk}	USL_{jk}	LSL_{jk}	B_{jk}	C_{jk}	A_{jk}
1	41.3	33.3	402	304	0.631	29	17	331	241	0.964	681	405	914	714	0.735
2	48.2	35.2	465	318	0.917	28	20	333	264	0.946	657	417	927	655	0.372
3	45.7	37.6	449	381	0.234	31	16	359	248	0.842	645	432	888	615	0.282
4	48.1	39.4	485	393	0.038	26	19	365	245	0.37	627	443	880	623	0.069
5	41.3	32.1	492	390	0.347	32	18	324	244	0.189	644	428	929	659	0.207
6	46.8	35.6	436	345	0.812	32	24	369	267	0.641	583	397	928	620	0.06
7	43.9	34.8	448	353	0.996	27	19	366	243	0.831	614	405	866	675	0.303
8	44.5	37.2	456	391	0.995	33	20	356	263	0.727	665	429	859	668	0.863
9	49.1	39.8	485	350	0.734	27	16	328	257	0.539	674	424	881	717	0.526
10	47.4	30.6	462	353	0.526	33	20	349	267	0.058	599	402	912	679	0.487

Note: Values are normalized average for all 50 industrial companies

TABLE 9.3
Investment, elasticity and workforces corresponding to each sustainability index

	Sustainability factors	Corresponding index				
		Index 1	Index 2	Index 3	Index 4	Index 5
L_{ik}	Environmental	11	14	13	17	12
	Social	7	6	5	3	8
	Economic	15	17	19	16	17
β_{ik}	Environmental	0.35	0.29	0.31	0.28	0.33
	Social	0.17	0.12	0.15	0.14	0.11
	Economic	0.47	0.56	0.68	0.51	0.78
κ_{ik}	Environmental	125	145	163	152	138
	Social	78	95	83	76	91
	Economic	265	248	319	296	311

TABLE 9.4
SPV of all companies

No.	1	2	3	4	5	6	7	8	9	10
SPV	0.635	0.269	0.456	0.759	0.233	0.568	0.654	0.852	0.745	0.562
No.	11	12	13	14	15	16	17	18	19	20
SPV	0.311	0.421	0.458	0.531	0.615	0.741	0.238	0.445	0.701	0.428
No.	21	22	23	24	25	26	27	28	29	30
SPV	0.444	0.215	0.560	0.128	0.711	0.802	0.903	0.774	0.625	0.447
No.	31	32	33	34	35	36	37	38	39	40
SPV	0.625	0.611	0.714	0.782	0.806	0.226	0.328	0.429	0.753	0.108
No.	41	42	43	44	45	46	47	48	49	50
SPV	0.405	0.336	0.289	0.417	0.198	0.608	0.708	0.337	0.469	0.851
Total SPV of industrial cluster					0.683					

own benefits. It has been estimated that by embracing and participating in the technologies, data available, and other elements of Industry 4.0, Quality 4.0, can potentially reduce the total cost of quality along with the practical quality management application and manufacturing in the industrial cluster. Mostly, managers decide to employ Industry 4.0 technologies and analyze their impacts on the sustainable quality management of the whole cluster. Gaining cluster support for investments in quality management and Industry 4.0 technologies is critical to success. It is essential to demonstrate how those technologies will improve overall quality. This process is elaborated using the sustainability concept. According to the proposed methodology, in the first step the Delphi method is used to collect Lean Six Sigma elements as an effective quality management paradigm. Then, a purification of Lean Six Sigma regarding sustainability main factors is performed using AISM. The results of these two steps are summarized in Table 9.1.

Those two steps are handled by a group of experts from the main management board of the industrial cluster. Their experiences and acquaintance with Industry 4.0 were sufficient to distinguish the necessity of implementation and the expected profit to be achieved.

Next, the quantification phase should be done to investigate the obtained production sustainability consequence from the main sustainability factors integrated with Lean Six Sigma elements. To quantify according to Equation (1), upper and lower specifications are collected from the quality assurance departments of all 50 companies of the industrial cluster. Costs and benefits of implementing each element and sustainability index were reported by the finance departments. Note that the 12 periods in one year are considered and an interest rate of 17 percent is announced by the central bank of Iran. Some input data are presented in Table 9.2.

Also, workforce and investment required for sustainability indices are given in Table 9.3.

Then, according to Equation (1), total sustainable production value (see Table 9.4) of the industrial cluster and also the SPV for each company are obtained.

The closer the SPV to 1, the more sustainable is the production cluster. Also, the total SPV for the industrial cluster is 0.683 showing that to have a more desirable sustainable production, sustainability factors and LSS elements should be fortified as discussed in the next section.

9.6 DISCUSSION AND IMPLICATIONS

The management board of an industrial cluster in the Northern part of Iran offers the complete value creation from the initial idea to the finished product. Based on the variety of different and customized products, the manufacturing environment is characterized by several stand-alone solutions which perform distributed and complex production steps. By a seamless integration and interconnection of all different production processes and steps, managers expect more transparency in its production processes coupled with less required effort. Through the usage of Industry 4.0 sensors at each production stage, the chance to reach a new level of identifying quality elements is considered. These elements are related to causes and issues which should reduce scraps and reworks. As a result of connecting different systems, new daily production and business metrics become available in real time which provide the necessary information to change and adjust processes or resources without delay through sustainable production. In addition, networked processes and machinery should enable a preventive action when undesirable events are approaching. As a result, responsible persons should immediately be notified about problems that are occurring in the manufacturing workflow and the roots of the problems should be easily identified based on the new collected data and information. Through the tight networking of all machines and products, quality related information about failures and problems can be shared between the machines so that the machines can immediately inform the operator. This would address the common problem of information sharing between people working on different shifts.

The main improvement of the integrated Lean Six Sigma sustainable value evaluation in the horizontal integration of the value chain is to better integrate its customers and suppliers. Today, many customers must call the company to get information about the current state of their ordered product. By a tight horizontal integration of all systems along the entire value chain, the customers are able to track the progress of their ordered product. Customers can get the current status of the product in the manufacturing process at any time and can see which tasks are already fulfilled and which are coming next. In case of problems, the customers can be immediately informed and provided with actual data in order to decide how to deal with the arising problem.

Moreover, today's complex products are often changed during the manufacturing process. Especially in the context of manufacturing electronic modules, where a diversity of complex components are mounted and soldered, changes occur frequently. Based on a close integration of the customer in the manufacturing process, changes and change requests can be handled in a better way and the customer is able to trigger the changes at the desired stage in the production process. In addition, the customer immediately gets feedback about the effects of changes and can regulate the quality based on the elements of Lean. Through a close cooperation, enabled by a horizontal integration of manufacturing systems with the quality management system, developers and engineers can analyze the behavior of their products at each production stage and are able to react very quickly to undesired outcomes through a sustainable approach. Beside these opportunities, customers also can profit in the context of complaints and required quality assessments. Feedback on complaints can be handled much faster and in much more detail. In addition, the managers can immediately monitor the countermeasures taken in the production process. The horizontal integration is not only beneficial to the production manager but, in addition, a high potential to optimize the flow of goods exists when suppliers get real time information about the actual stocks of in-house components and future demands. Therefore, suppliers can better plan and organize the delivery and immediately share information about components that are not available any more. Due to the immense variety of different components on today's complex electronic modules and devices, many suppliers can deliver the necessary number of components for their products. A networked integration of different suppliers increases the chance of reducing missing parts and of avoiding tying up any money in stocks where production processes typically only start if all necessary components have arrived at the factory. Moreover, suppliers and manufacturers of components get data directly on how their components and parts perform in the manufacturing processes (for example, defects and the waste of components because of the applied solder temperature profile). This enables suppliers to optimize their product portfolio and manufacturers to improve their components in accordance with real manufacturing data.

By an end-to-end digital integration of the sustainable production system, managers consider the opportunity to reduce the currently huge amount of data transformations along the life cycle of a product. The digital incorporation of each stage of a product's life cycle enables new synergies and opportunities to optimize sustainability along the entire value chain of the product. Products can carry relevant information and data about themselves and can provide this information at each stage of their life cycle. Furthermore, managers expect a reduction of time to market of new products if engineers and developers can use complete digital models in a sustainable process. Therefore, they are able to foresee the results and effects of their product designs.

9.7 CONCLUSIONS

Summarizing, Industry 4.0 provides a huge potential through the different concepts in production systems. Through implementing concepts of Industry 4.0, the industrial cluster in this study considered the chance to fundamentally improve their entire production planning, optimize the flow of goods, optimize the quality of processes and products, and strengthen the customer and supplier relationships as well as the potential for offering new business models through a comprehensive sustainable production paradigm. Moreover, decision makers expect a further automation of processes and a high level of digitalization, which should make the lives of the employees easier and reduce their workload. In addition, policy makers anticipate a noteworthy decrease in the entire manufacturing lead-time and simultaneously a reduction of buffer stock as well as the elimination of overproduction through implementing an Industry 4.0 driven decentralized and self-organized manufacturing structure. Apart from these chances, authors think the main upcoming challenges are filtering and processing the huge amount of new information, which is provided by Industry 4.0 in order to avoid information overload. A further challenge is the determination of the defined rules in which the autonomous smart machines and systems can decide on their own. Therefore, the range of freedom for each smart device must be carefully determined to ensure compliance with the regulations and leanness of the production sustainability. Further, methods to measure the efficiency and effectiveness of each process must be established and applied. Moreover, means should be determined to prevent nonconformities and eliminate their causes. Lastly, a process of continual improvement of the quality management system should be established and applied. It is assumed that the proposed quantification formulation improves a quality management system and is an appropriate means to construct a link between Industry 4.0 and the quality management domain since they are neither too abstract nor too detailed to act as a basis for configuring a comprehensive sustainable production system.

BIBLIOGRAPHY

Alayon, C., Safsten, K., & Johansson, G. (2017). Conceptual sustainable production principles in practice: do they reflect what companies do?. *Journal of Cleaner Production,* 141, 693–701.

Artiach, T., Lee, D., Nelson, D., et al (2010). The determinants of corporate sustainability performance. *Accounting & Finance*, 50(1), 31–51.

Badurdeen, F., & Jawahi, I.S. (2017). Strategies for value creation through sustainable manufacturing. Procedia Manufacturing, 8, 20–27.

Bag, S., Yadav, G., Dhamija, P., et al. (2021). Key resources for Industry 4.0 adoption and its effect on sustainable production and circular economy: An empirical study. *Journal of Cleaner Production*, 281, 125233.

Banawi, A., & Bilec, MM. (2014). A framework to improve construction processes: Integrating Lean, Green and Lean Six Sigma. *International Journal of Construction Management*, 14(1), 45–55.

Bastas, A., & Liyanage, K. (2018). Sustainable supply chain quality management: A systematic review. *Journal of Cleaner Production*, 181, 726–744.

Bernal, E., Edgar, D., & Burnes, B. (2018). Building sustainability on deep values through mindfulness nurturing. *Ecological Economics*, 149, 645–657.

Bhuiyan N., & Baghel, A. (2005) "An overview of continuous improvement: from the past to the present. *Management Decision*, 43(5), 761–771.

Cai, W., Liu, C., Zhang, C., et al. (2018). Developing the ecological compensation criterion of industrial solid waste based on energy for sustainable development. *Energy*, 157, 940–948.

Cai, W., Liu, F., Hu S. (2017) An analytical investigation on energy efficiency of high-speed dry-cutting CNC hobbling machines. *International Journal of Sustainable Engineering*, *(11)*6, 412–419.

Calia, R.C., Guerrini, F.M., & Castro, M de. (2009). The impact of Lean Six Sigma in the performance of a pollution prevention program. *Journal of Cleaner Production*, 17(15), 1303–1310.

Chen, B., Wan, J., Shu, L. et al. (2018). Smart Factory of Industry 4.0: Key technologies, application case, and challenges. *IEEE Access*, 6, 6505–6519.

Cherrafi, A., Elfezazi, S., Chiarini, A., et al. (2016). The integration of lean manufacturing, Lean Six Sigma and sustainability: A literature review and future research directions for developing a specific model. *Journal of Cleaner Production*, 139, 828–846.

Cluzel, F., Yannou, B., Afonso, D., et al. (2010). Managing the complexity of environmental assessments of complex industrial systems with a Lean 6 Sigma. *Complex Systems Design & Management*, 279–294.

Despeisse, M., Mbaye, F., Ball, P.D., et al. (2012). The emergence of sustainable manufacturing practices. *Production Planning & Control*, 23(5), 354–376.

EPA. 2006. The Lean and Environment Toolkit. Accessed July 10, 2015. www.epa.gov/Lean/environment/toolkits/index.htm

Erdil, N.O., Aktas, C.B., & Arani, O.M.(2018). Embedding sustainability in Lean Six Sigma efforts. *Journal of Cleaner Production*, 198, 520–529.

Fazlollahtabar, H. (2016). Parallel autonomous guided vehicle assembly line for a semi-continuous manufacturing system. *Assembly Automation*, 36(3), 262–273.

Fazlollahtabar, H. (2018a). Lagrangian relaxation method for optimizing delay of multiple autonomous guided vehicles. *Transportation Letters*, 10(6), 354–360.

Fazlollahtabar, H. (2018b). Scheduling of multiple autonomous guided vehicles for an assembly line using minimum cost network flow. *Journal of Optimization in Industrial Engineering*, 11(1), 185–193.

Fazlollahtabar, H. (2019a). An effective mathematical programming model for production of automatic robot path planning. *The Open Transportation Journal*, 13(1), 11–16.

Fazlollahtabar, H. (2019b). Triple state reliability measurement for a complex autonomous robot system based on extended triangular distribution. *Measurement*, 139, 122–126.

Fazlollahtabar, H. (2020). Comparative simulation study for configuring turning point in multiple robot path planning: Robust data envelopment analysis. *Robotica*, 38(5), 925–939.

Fazlollahtabar, H. (2021). Robotic Manufacturing Systems Using Internet of Things: New Era of Facing Pandemics. *Automation, Robotics & Communications for Industry*, 4.0, 82.

Fazlollahtabar, H. (2022). Internet of Things-based SCADA system for configuring/reconfiguring an autonomous assembly process. *Robotica*, 40(3), 672–689.

Fazlollahtabar, H., & Hassanli, S. (2018). Hybrid cost and time path planning for multiple autonomous guided vehicles. *Applied Intelligence,* 48, 482–498.

Fazlollahtabar, H., & Jalali, S.G. (2013). Adapted Markovian model to control reliability assessment in multiple AGV. *Scientia Iranica,* 20(6), 2224–2237.

Fazlollahtabar, H., Mahdavi-Amiri, N., & Muhammadzadeh, A. (2015). A genetic optimization algorithm for nonlinear stochastic programs in an automated manufacturing system. *Journal of Intelligent & Fuzzy Systems,* 28(3), 1461–1475.

Fazlollahtabar, H., & Niaki, S.T.A. (2017a). Binary state reliability computation for a complex system based on extended Bernoulli trials: Multiple autonomous robots. *Quality and Reliability Engineering International,* 33(8), 1709–1718.

Fazlollahtabar, H., & Niaki, S.T.A. (2017b). Integration of fault tree analysis, reliability block diagram and hazard decision tree for industrial robot reliability evaluation. *Industrial Robot: An International Journal,* 44(6), 754–764.

Fazlollahtabar, H., & Niaki, S.T.A. (2017c). *Reliability Models of Complex Systems for Robots and Automation.* CRC Press.

Fazlollahtabar, H., & Niaki, S.T.A. (2018a). Cold standby renewal process integrated with environmental factor effects for reliability evaluation of multiple autonomous robot system. *International Journal of Quality & Reliability Management,* 35(10), 2450–2464.

Fazlollahtabar, H., & Niaki, S.T.A. (2018b). Modified branching process for the reliability analysis of complex systems: Multiple-robot systems. *Communications in Statistics-Theory and Methods,* 47(7), 1641–1652.

Fazlollahtabar, H., & Saidi-Mehrabad, M. (2015a). *Autonomous Guided Vehicles: Methods and Models for Optimal Path Planning.* Springer International Publishing, Germany.

Fazlollahtabar, H., & Saidi-Mehrabad, M. (2015b). Risk assessment for multiple automated guided vehicle manufacturing network. *Robotics and Autonomous Systems,* 74, 175–183.

Fazlollahtabar, H., & Saidi-Mehrabad, M. (2019). *Cost Engineering and Pricing in Autonomous Manufacturing Systems.* Emerald Publishing Limited.

Fazlollahtabar, H., Saidi-Mehrabad, M., & Balakrishnan, J. (2015a). Mathematical optimization for earliness/tardiness minimization in a multiple automated guided vehicle manufacturing system via integrated heuristic algorithms. *Robotics and Autonomous Systems,* 72, 131–138.

Fazlollahtabar, H., Saidi-Mehrabad, M., & Balakrishnan, J. (2015b). Integrated Markov-neural reliability computation method: A case for multiple automated guided vehicle system. *Reliability Engineering & System Safety,* 135, 34–44.

Fazlollahtabar, H., Saidi-Mehrabad, M., & Masehian, E. (2015). Mathematical model for deadlock resolution in multiple AGV scheduling and routing network: A case study. *Industrial Robot: An International Journal,* 42(3), 252–263.

Fazlollahtabar, H., Saidi-Mehrabad, M., & Masehian, E. (2021). Robotic industrial automation simulation-optimization for resolving conflict and deadlock. *Assembly Automation,* 41(4), 477–485.

Fazlollahtabar, H., & Shafieian, S.H. (2014). An optimal path in an AGV-based manufacturing system with intelligent agents. *Journal for Manufacturing Science and Production,* 14(2), 87–102.

Feng, L., Mears, L., Beaufort, C., et al. (2016). Energy, economy, and environment analysis and optimization on manufacturing plant energy supply system. *Energy Convers Manage,* 117, 454–65.

Figge, F., & Hahn, T. (2004). Sustainable Value Added—measuring corporate contributions to sustainability beyond eco-efficiency. *Ecological Economics* 48(2), 173–187.

Freitas, J.G de. Costa, H.G., & Ferraz, F.T. (2017). Impacts of Lean Six Sigma over organizational sustainability: A survey study. *Journal of Cleaner Production,* 156, 262–275.

Gabriel, M., & Pessel, E. (2016). Industry 4.0 and sustainability impacts: Critical discussion of sustainability aspects with a special focus on future of work and ecological consequences. *International Journal of Engineering*, 1, 131–136.

Garvare, R., & Johansson, P. (2010). Management for sustainability–A stakeholder theory. *Total Quality Management and Business Excellence*, 21(7), 737–744.

Garza-Reyes, J.A. (2015). Lean and green–a systematic review of the state of the art literature. *Journal of Cleaner Production*, 102, 18–29.

Hajmohammad, S., Vachon, S., Klassen, R.D., et al. (2013). Reprint of Lean management and supply management: Their role in green practices and performance. *Journal of Cleaner Production*, 56, *86–93*.

Han, S.H., Chae, M.J., Im, K.S. et al (2008). Lean Six Sigma–based approach to improve performance in construction operations. *Journal of Management in Engineering*, 1(24), 21–231.

Hartini, S., & Ciptomulyono, U. (2015). The Relationship between Lean and Sustainable Manufacturing on Performance. *Literature Review. Procedia Manufacturing*, 4, 38–45.

Hassini, E., Surti, C., & Searcy, C. (2012). A literature review and a case study of sustainable supply chains with a focus on metrics *International Journal of Production Economics*, 140(1), 69–82.

Henriques, J., & Catarino, J. (2015). Sustainable value and cleaner production–research and application in 19 Portuguese SME. *Journal of Cleaner Production*, 96, 379–386.

Johansson, G., & Sundin, E. (2013). Lean and Green Product Development: Two Sides of the Same Coin?. *Journal of Cleaner Production*, 85, 104–121.

Liao, Y., Loures E.d.F.R., & Deschamps, F. (2018). Industrial Internet of Things: A systematic literature review and insights. *IEEE Internet of Things Journal,* vol. 5, no 6, pp. 4515-4525.

Linderman, K., Schroeder, R., Zaheer, S., et al. (2003). Lean Six Sigma: A goal–theoretic perspective. *Journal of Operation Management*, 21(2), 193–203.

Lv, J., Peng, T., & Tang, R. (2018). Energy modeling and a method for reducing energy loss due to cutting load during machining operations. *The Journal of Engineering Manufacture*, 233(3), 699–710.

Ma, M., & Cai, W. (2018a). What drives the carbon mitigation in Chinese commercial building sector? Evidence from decomposing an extended Kaya identity. *Science of The Total Environment*, (634), 884–899.

Ma, M., & Cai, W. (2018b). Do commercial building sector-derived carbon emissions decouple from the economic growth in Tertiary Industry? A case study of four municipalities in China. *Science of the Total Environment*, 650, 822–834.

Mathai, M.V., Isenhour, C., Stevis, D., et al. (2021). The political economy of (Un)sustainable production and consumption: A multidisciplinary synthesis for research and action, resources, *Conservation and Recycling*, 167, 105265.

Mayer, F., & Pantförder, D. (2014). Unterstützung des Menschen in cyber-physical-production systems. In: Bauernhansl, T., ten Hompel, M., Vogel-Heuser, B. (eds.) *Industrie 4.0 in Produktion, Automatisierung und Logistik*, pp. 481–491. Springer Vieweg, Wiesbaden.

Mikulčić, H., Vujanović, M., & Duić N. (2013). Reducing the CO2 emissions in Croatian cement industry. *Applied Energy*, 101, 41–48.

Miller, G., Pawloski, J., & Standridge, C.R. (2010). A case study of lean, sustainable manufacturing. *Journal of Industrial Engineering and Management*, 3(1), 11–32.

Pampanelli, A.B. Found, A.B., & Bernardes, A.M. (2014). A Lean & Green Model for a production cell. *Journal of Cleaner Production*, 85, 19–30.

Putnik, D.G., Varela, R.L., Carvalho, C., et al. (2015). Smart objects embedded production and quality management functions. *International Journal for Quality Research* 9, 151–166.

Rauch, E., Dallasega, P., & Matt, DT. (2016). Sustainable production in emerging markets through Distributed Manufacturing Systems (DMS). *Journal of Cleaner Production*, 135, 127–138.

Resta, B., Dotti, S., Gaiardelli, P. et al. (2016). Lean manufacturing and sustainability: An integrated view. *IFIP International Conference on Advances in Production Management Systems*, 569–666.

Rosen, M.A., & Kishawy, H.A. (2012a). Sustainable manufacturing and design: Concepts, practices and needs. *Sustainability*, 2(4), 154–174.

Rosen, M.A., & Kishawy, H.A. (2012b). Sustainable manufacturing and design: Concepts, practices and needs. *Sustainability*, 4(2), 154–174.

Sharrard, A.L. Matthews, H.S., & Ries, R.J. (2008). Estimating construction project environmental effects using an input-output-based hybrid life-cycle assessment model. *Journal of Infrastructure Systems*, 14(4), 327–336.

Shojaeifar, A., Fazlollahtabar, H., & Mahdavi, I. (2016). Decomposition versus minimal path and cuts methods for reliability evaluation of an advanced robotic production system. *Journal of Automation Mobile Robotics and Intelligent Systems*, 10(3), 52–57.

Shrivastava, P., & Berger. (2010). Sustainability principles: A review and directions. *Management & Organization*, 7(4), 246–261.

Solding, P., Petku, D., & Mardan, N. (2009). Using simulation for more sustainable production systems. *International Journal of Sustainable Engineering*, 2(2), 111–122.

Tao, J., & Yu, S.H. (2018). Product life cycle design for sustainable value creation: Methods of sustainable product development in the context of high value engineering. *Procedia CIRP*, 69, 25–30.

Veleva, V., & Ellenbecker, M. (2001). Indicators of sustainable production: Framework and methodology. *Journal of Cleaner Production*, 9(6), 519–549.

Wojnarowska, M., Sołtysik, M., & Prusak, A. (2021). Impact of eco-labelling on the implementation of sustainable production and consumption, *Environmental Impact Assessment Review*, 86, 106505.

Zhang, Z., & Awasthi, A. (2014). Modelling customer and technical requirements for sustainable supply chain planning. *International Journal of Production Research*, 52(17), 5131–5154.

10 Optimization Models

10.1 INTRODUCTION

Scheduling, as a decision process, plays a key role in most manufacturing and production systems, in transportation and in the other kinds of industries. The utilization of robots in the manufacturing industries leads to a variety of advantages ranging from preciseness in parts processing to high volume of productivity. Generally, these operations are required to be done on all parts in the same order, and in the most cases, there is more than one processing machine in each stage. This environment is referred to as a flexible flow-shop. A robotic flexible-flow-shop contains a series of robots (AGVs) that are accountable for conveying parts between pairs of consecutive stations (machines), which is common in automatic manufacturing systems. Due to growing customers' needs for more products, and requirements for more precisely manufactured products, it is of significance for all manufacturers. In order to keep up with pioneer industries, more underline the robotic-services' roles in production. In the first case, we study a multi-stage robotic-flexible-flow-shop having a number of AGVs whose assignments are to convey parts onto all machines existing in the stages, and there are sequence-dependent setup times for all parts.

Kalczynski and Kamburowski (2012) showed that a two-machine flow-shop with release times and an objective function of makespan is strongly NP-hard. Sangsawang et al. (2015) considered the two-stage Reentrant Flexible-Flow-Shop (RFFS) with blocking constraint, whose objective is minimizing the makespan. They applied the Hybrid Genetic Algorithm (HGA) with adaptive auto-tuning based on a fuzzy logic controller and Hybrid Particle Swarm Optimization (HPSO) with Cauchy distribution to solve the problem. Moslehi and Khorasanian (2013) used the blocking flow shop scheduling problem for minimizing the total completion time criterion. They introduced two mixed binary integer-programming models, the first one is modeled based on the departure times of jobs from machines, and the second is modeled based on the idle and blocking times of jobs. Wang et al. (2011) to solve the blocking permutation flow shop scheduling problem with total flow time criterion, proposed a Hybrid Modified Global-Best Harmony Search (hmgHS) algorithm. Ruiz and Rodriguez (2010) raised the idea that the properties of graph representation in the hybrid-flow-shop scheduling problem, suggested that critical block theory could be applied to accelerate the search procedure. Batur et al. (2010) optimized the cycle time

DOI: 10.1201/9781003400585-10

by two-stage heuristic for the scheduling problem in which there are two-machine robotic cells producing multiple parts. Lei (2015) considered a flow shop-scheduling problem with two objective functions and its feasibility was reviewed. The general purpose of the problem is the minimization of the first objective function and the total tardiness of the second objective function concurrently. Gaunlong et al. (2012) introduced the blocking flow shop scheduling problem and they used a discrete artificial bee colony algorithm to solve it. Sawik (2012) considered the flexible-flow-shop with finite in-process buffers, and continuous or limited machine availability. Elmi and Topaloglu (2013) dealt with hybrid flow shop scheduling with multiple robots. They proposed a mixed integer linear programming model, which minimized the makespan, and as a solution, a simulated annealing-based heuristic was introduced. Jabbarizadeh et al. (2009) considered hybrid flexible-flow-shops. They exerted sequence-dependent setup times and machine availability constraints. Behnamian and Zandieh (2011) minimized earliness and quadratic tardiness in the hybrid flow shop scheduling problem. They assumed that each stage minimally has one machine, and at least one of stages possesses more than one machine. Furthermore, there is no release time for jobs, which are independent, and each job must be allocated at most to one machine within one stage

10.2   OPTIMIZATION MODEL CASE I: ROBOTIC FLEXIBLE SCHEDULING PROBLEM

In this study, a robotic flexible flow-shop system is considered which consists of p parts that should be processed at s stages and all the p parts should pass all stages. In addition, m is attributed to stations (machines) within stages, except stations located in the first stage, which is the storehouse. The rest of the stages incorporate M_s stations $\left(M_s \geq 1 \right)$, and each station contains parallel processing machines that are supplied by AGVs. All tasks on machines such as loading, unloading and setup are performed by robotic-adaptive-grippers. Each part, which is transferred by one of the AGVs into each stage should be processed on one machine and consequently, each part should enter one stage by one AGV on one of its travels. Each AGV can only transfer one part in each travel, and cannot start another travel unless it has returned from previous travel. There is an unlimited buffer for each processing machine. The part should be waiting on machines until one of the AGVs comes and conveys the part to the next stage. In this case, the model imposes the setup time on two successive parts from different part families. We also consider travel time for AGVs, loading and unloading time, which is aggregated for each travel. The objective is to schedule all parts while minimizing the makespan. One of the most controversial issues which comes with the notation of AGV, is how to delineate a routing approach for AGVs since they may encounter one another while transferring parts. We assume that there are two separate routes between each pair of machines, one for going forward and one for going backward. To cope with this problem, we depict two accessibility pattern for AGVs to access the machines. We classify our problem with accordance to the often used three-field notation for machines scheduling problems $\alpha|\beta|\gamma$ of Graham et al. (1978), in which α describes the machine structure, β characterizes the part features and limitations, and γ delineates the objective to minimize. We apply the proposal introduced by Lee and

Chen (2001), and we show our problem as: $RFFS, m \in M | p \in P, a \in A, t \in T | C_{Max}$. In the α section, we use $RFFS, m \in M$ to denote a robotic-flexible-flow-shop problem with sequence–dependent setup time and AGV-based transportation system with overall m processing machines, In the section β, p stands for the number of parts, a demonstrates the number of AGVs, and t shows the number of travels for each AGV. In the section γ, C_{Max} means that we are going to minimize the makespan.

10.3 MATHEMATICAL FORMULATIONS

In this section, a Mixed Integer Linear Program (MILP) is formulated for a hybrid-flow-shop problem with Robotic Adaptive Gripper-based processing and AGV-based transportation system to minimize makespan.

10.3.1 ASSUMPTIONS

- There is at least one machine at each stage ($m \geq 1$).
- The capacity of each AGV for each travel is just one.
- All the required parts to be processed are entirely provided in storage.
- All parts should pass all stages.
- There are unlimited buffers for each machine.
- The sum of the loading and unloading time for each part is shown by one parameter (LT_p).

10.3.2 INDICES

P: Set of parts where,
p: Index of part, $p \in \{1, 2, \cdots, P\}$
F: Set of parts families where,
f: Index of parts family, $f \in \{1, 2, \cdots, F\}$
S: Set of stages where,
s: Index of stages, $s \in \{1, 2, \cdots, S\}$
e: Index of last stage
A: Set of AGVs where,
a: Index of AGVs, $a \in \{1, 2, \cdots, A\}$
T: Number of travels for AGVs where,
t: Index of travels, $t \in \{1, 2, \cdots, T\}$
M: Set of machines with Robotic Adaptive Grippers where,
m: Index of machines with Robotic Adaptive Grippers, $m \in \{1, 2, ..., M\}$.
M_s: Set of machines in stage s.

10.3.3 PARAMETERS

$ST_{p,m}$:   Required setup time for part p on machine m
$PT_{p,m}$:   Required time for processing part p on machine m
$D_{m',m}$:   Required time for AGV a to travel from machine m' to machine m
LT_p: Sum of the required time for Robotic Adaptive Grippers to load and unload each part p on AGV

X: Number of all required operations at the last stage
M: A large number
$R_{m',m,s}$:   Showing if AGV a is at machine m', to which next machine m at the stage s is permitted to go (it prevents assigning machines to stages irrationally)
$Q_{f,p}$:   Showing which parts p are included to part family f
$PA_{p,m}$:   Process ability, showing which machines m can process part p.

10.3.4 Variables

$RT_{m',m,a,t}$ The reception time of AGV's loads transferred during AGV's t^{th} travel from machine m' (located in stage s-1) to machine m (located in stage s)
$PC_{p,m',m,s}$ The process completion time of part p on machine m which is transferred from machine m' (located in stage s-1) to machine m (located in stage s)
$C_{Max}=$ The maximum completion time of all parts, makespan
$W_{m',m,a,t}=1$ if AGV a, in its tth travel goes from machine m' (located in stage s-1) to machine m (located in stage s), and 0 otherwise
$Y_{p,m',m,s}=1$ if part p is transferred from machine m' (located in stage s-1) to machine m which is located in stage s, and 0 otherwise
$SS_{p,p',s}=1$ if the part p is processed before the part p' in stage s, and 0 otherwise.

10.3.5 Mathematical Model

$$Minimize\, C_{Max}$$

Such that

$$\sum_{m'}^{M}\sum_{m}^{M}\sum_{p}^{P}\sum_{s}^{S} W_{p,m',m,a,t} \times R_{m',m,s} \times PA_{p,m} \leq 1 \forall a \in A, t \in T \qquad (1)$$

$$\sum_{m'}^{M}\sum_{m}^{M}\sum_{p}^{P}\sum_{s}^{S} W_{p,m',m,a,t} \times R_{m',m,s} \times PA_{p,m} \leq 1 \forall p \in P, s \in S, s > 1 \qquad (2)$$

$$\sum_{s}^{S} W_{p,m',m,a,t} \times R_{m',m,s} \times H \geq RT_{p,m',m,a,t} \quad \forall p \in P, m \in M, m' \in M, a \in A, t \in T \qquad (3)$$

$$Y_{p,m',m,s} \times H \geq PC_{p,m',m,s} \quad \forall p \in P,, m \in M, m' \in M, s \in S, s > 1 \qquad (4)$$

$$Y_{p,m',m,s} \leq \sum_{a}^{A}\sum_{t}^{T} W_{p,m',m,a,t} \quad \forall p \in P, m \in M, m' \in M, s \in S, s > 1 \qquad (5)$$

$$H \times (2 - \sum_{m'}^{M}\sum_{p'}^{P} W_{p',m',m,a,t-1} - \sum_{m'}^{M} W_{p,m'',m',a,t}) + \sum_{m'}^{M} RT_{p,m'',m',a,t} \geq$$
$$\sum_{m'}^{M}\sum_{p'}^{P} RT_{p',m',m,a,t-1} + D_{m,m''} + LT_p + \sum_{m'}^{M} W_{p,m'',m',a,t} \times D_{m'',m'} \qquad (6)$$
$$\forall a \in A, t \in T, m \in M, m'' \in M, p \in P$$

$$\sum_{m}^{M}\sum_{m'}^{M}\sum_{p}^{P}W_{p,m',m,a,t-1} \geq \sum_{m}^{M}\sum_{m'}^{M}\sum_{p}^{P}W_{p,m',m,a,t} \quad \forall a \in A, t \in T, t > 1 \qquad (7)$$

$$H \times (1 - \sum_{m}^{M}W_{p,m',m,a,t}) + \sum_{m}^{M}RT_{p,m',m,a,t} \geq \sum_{m}^{M}PC_{p,m,m',s-1} + LT_p$$
$$+ \sum_{m}^{M}W_{p,m',m,a,t} \times D_{m',m} \; \forall p \in P, a \in A, t \in T, m' \in M, s \in S, s > 1 \qquad (8)$$

$$H \times (1 - \sum_{m'}^{M}Y_{p,m',m,s}) + \sum_{m'}^{M}PC_{p,m',m,s} \geq \sum_{m'}^{M}RT_{p,m',m,a,t} + PT_{p,m} \qquad (9)$$
$$\forall p \in P, a \in A, t \in T, m \in M, s \in S, s > 1$$

$$\sum_{m'}^{M}\sum_{m}^{M}\sum_{p}^{P}Y_{p,m',m,s} = X \quad \forall s \in S, s = e \qquad (10)$$

$$H \times (2 - \sum_{m'}^{M}Y_{p,m',m,s} \times Q_{f,p} - \sum_{m'}^{M}Y_{p',m',m,s} \times Q_{f,p'}) + (1 - ss_{p,p',m}) \times H$$
$$+ \sum_{m'}^{M}PC_{p,m',m,s} \geq \sum_{m'}^{M}PC_{p',m',m,s} + PT_{p,m} \qquad (11)$$
$$\forall p, p' \in P, p \neq p', m \in M, f \in F, s \in S, s > 1$$

$$H \times (2 - \sum_{m'}^{M}Y_{p,m',m,s} \times Q_{f,p} - \sum_{m'}^{M}Y_{p',m',m,s} \times Q_{f,p'}) + (ss_{p,p',m}) \times H$$
$$+ \sum_{m'}^{M}PC_{p',m',m,s} \geq \sum_{m'}^{M}PC_{p,m',m,s} + PT_{p',m} \qquad (12)$$
$$\forall p, p' \in P, p \neq p', m \in M, f \in F, s \in S, s > 1$$

$$H \times (2 - \sum_{m'}^{M}Y_{p,m',m,s} \times Q_{f,p} - \sum_{m'}^{M}Y_{p',m',m,s} \times Q_{f',p'}) + (ss_{p,p',m}) \times H$$
$$+ \sum_{m'}^{M}PC_{p,m',m,s} \geq \sum_{m'}^{M}PC_{p',m',m,s} + PT_{p,m} + ST_{p,m} \qquad (13)$$
$$\forall p, p' \in P, p \neq p', m \in M, f, f' \in F, f \neq f', s \in S, s > 1$$

$$H \times (2 - \sum_{m'}^{M}Y_{p,m',m,s} \times Q_{f,p} - \sum_{m'}^{M}Y_{p',m',m,s} \times Q_{f',p'}) + (ss_{p,p',m}) \times H$$
$$+ \sum_{m'}^{M}PC_{p',m',m,s} \geq \sum_{m'}^{M}PC_{p,m',m,s} + PT_{p',m} + ST_{p',m} \qquad (14)$$
$$\forall p, p' \in P, p \neq p', m \in M, f, f' \in F, f \neq f', s \in S, s > 1$$

$$\sum_{m'}^{M} Y_{p,m',m,s} = \sum_{m'}^{M} Y_{p,m,m',s+1} \ \forall m \in M, p \in P, s \in S, 1,< s < e \qquad (15)$$

$$\sum_{m'}^{M} \sum_{m}^{M} PC_{p,m',m,s} \leq C_{max} \ \forall p \in P, s \in S, s = e \qquad (16)$$

$$W_{p,m',m,a,t} \in \{0,1\} \ \forall p \in P, m, m' \in M, a \in A, s \in S, s < 1 \qquad (17)$$

$$Y_{p,m',m,s} \in \{0,1\} \ \forall p \in P, m, m' \in M, s \in S, s < 1 \qquad (18)$$

$$SS_{p,p',m} \in \{0,1\} \ \forall p, p' \in P, m \in M \qquad (19)$$

$$RT_{p,m',m,a,t} \geq 0 \forall p \in P, m, m' \in M, a \in A, s \in S, s < 1 \qquad (20)$$

$$PC_{p,m',m,s} \geq 0 \forall p \in P, m, m' \in M, s \in S, s < 1 \qquad (21)$$

Constraint (1) states that each AGV can only be allocated to one route on each of its travels. Constraint (2) shows that each part can be processed on just one machine at each stage. Constraint (3) suggests that the Reception Time (RT), to take a non-zero value, needs to have binary variable W taken as 1. Constraint (4) implies that the Process Completion time (PC), to take a non-zero value, needs to have binary variable Y taken as 1. Constraint (5) maintains that first, a part should be transferred to one machine, then that part can starts its process on that machine. Constraint (6) states that Reception Time (RT) of non-first travel of an AGV, should be equal or more than the sum of the previous travel's Reception Time (RT), the time required to move from the previous travel machine to a new machine to load the particular part, the required time to load the particular part, and the time required to move from the new machine to the ultimate machine on its current travel. Constraint (7) represents that one AGV can start its non-first travel, when it has finished its previous travel. Constraint (8) mentions that the Reception Time of a part (RT) should be equal to or more than the sum of Process Completion times of that part (PC) on a machine in previous stages, and Loading Time of that part (LT), and the time required to travel from that machine in a previous stage to the current machine in the current stage. Constraint (9) expresses that the Process Completion time of a part (PC) should be equal to or more than the sum of Reception Times (RT), and Process Times (PT). Constraint (10) puts forward that the number of parts on the last stage should be equal to that of the first stage. Constraints (11) and (12) show process priority for parts at a certain machine where parts are from an identical part family $(f \in F)$. Constraints (13) and (14) illustrate process priority for parts at a certain machine where parts are not from an identical parts family $(f, f' \in F, f \neq f')$ (part families are different). Constraint (15) expresses that each part that enters one machine should exit that machine, unless that machine is located in the first or last stage (the first machine (storage) does not have input, and the last machine does not have output). Constraint (16) implies the maximum completion time (makespan). Constraint (17) to (21) shows non-negative and binary variables.

10.4 IMPLEMENTATION

Mostly, scheduling and routing problems are NP-hard, meaning that one problem including both these concepts as our problem, is considered completely NP-hard. Hence, in semi-large-scaled examples, the problem cannot be polynomially solved by an exact method such as CPLEX. To deal with this problem, we rule out some complexities of the model and limit search area in such a way that the ultimate output does not change. The approaches to limit the search area is:

1. Restricting AGVs to certain machines. In other words, not all the AGVs can visit all machines as they could in the normal version of the problem.
2. Limiting parts to certain machines. For example, machines 2 and 3 can process parts A and B. We allocate part A to machine 2, and part B to machine 3.

S1 to S4 shows the number of stages in which M1 to M5 machines exist (M1 is storage). P1 to P3 stand for the number of parts, and A1 to A2 are the number of AGVs, and T1 to T5 are considered the number of travels for each AGV. The upper side of the chart shows the movements of parts on machines, and the lower side of the chart exhibits the movements of the AGVs. To simplify the tracking of materials in the process, we allocated distinguishable amounts to parameters. Processing time and setup time are equal for each part (P1 = 6 for both of setup and process time, P2 = 7, and P3 = 8), and the time required for each part to be loaded on an AGV is: P1 = 1, P2 = 2, and P3 = 3. The distance between each pair of machines is shown in Table 10.1.

To better understand the tracking of the process using Table 10.1, we assess one part of AGV movement. AGV1 transfers P2 on its first travel from M1(storage) to M2, then moves to M3 to transfer P1 from machine M3 to M4 on its second travel (T2). It moves from machine M4 to machine M2 to transfer P2 from M2 to M4 on its third travel (T3), then it carries P1 from M4 to M5 on its fourth travel (T4). Finally, it goes back from M5 to M4 to load P2, and moves again from M4 to M5 on its fifth travel (T5). For example, on AGV1's first trip, there are two units of time between M1 to M2, and two units of time for loading P2. Therefore, delivery time for AGV1's first trip is 4. Then, it moves from M2 to M3 which takes 1 minute to travel. AGV1 reaches machine M3 at the time 5, and waits 4 minutes until P1 gets ready to travel (the

TABLE 10.1
Distance between machines (time)

Distance between machines	Station 1 (Storage)	Machine 2	Machine 3	Machine 4	Machine 5
Station 1(Storage)	—	2	2	4	6
Machine 2	2	—	1	2	4
Machine 3	2	1	—	2	4
Machine 4	4	2	2	—	2
Machine 5	6	4	4	2	—

ending process is 9). P1 takes 1 minute to be loaded on AGV1, and takes 2 minutes to be transferred from M3 to M4. As we expect, P1 reaches M4 at a time of 12.

In the normal condition, in which the model can allocate all the parts to all the AGVs, and machines, it takes 3518 seconds to reach 54 objective functions, which in comparison with the model's scale; it is too much time to solve. In order to apply those approaches, we assign P1 and P2 to be processed on M2 in stage 2, and P3 to be processed on M3 in stage 2. Secondly, A1 can transfer P1 and P2, and A2 can transfer P1 and P3. Therefore, this way the unnecessary search area is reduced, and the model can reach the desirable objective function of 54 in a time of just 7 seconds instead of 3518.

10.5 OPTIMIZATION MODEL CASE II: CLUSTERING-BASED UNCERTAIN SCHEDULING

10.5.1 HYBRID FLOW SHOP SCOPE

Scheduling is a form of decision-making process that plays a pivotal role in manufacturing and service systems. It is a process of allocating available and limited resources such as machines, material handling equipment, operators and tools over a specific time to perform a set of required tasks in order to achieve certain objectives. Hybrid Flow Shop (HFS) is an extension of two particular types of scheduling problems: Parallel Machine Scheduling (PMS) and Flow Shop Scheduling (FSS). It is a common manufacturing environment, which consists of a set of two or more processing stages with at least one stage having two or more parallel machines. The duplication of machines at some stages can introduce flexibility, increase capacities and avoid bottlenecks. HFS has been applied in many industries such as the chemical industry, automobile, the steel industry, spring, wire, semiconductor and electronics manufacturing. Thus, the HFS problem has a very great importance and has attracted much attention from researchers. In HFS problems, it is assumed that a set of N jobs must be sequentially processed at a set of k production stages or machine centers, each of which has several machines operating in parallel. Each stage has m_i, $i=1,2$, ..., k number of parallel processors. It is convenient to view a job as a sequence of k tasks, one task for each stage, where the processing of each task can be started whenever the preceding task is completed. Machines at each stage may be identical, uniform or unrelated and each machine within each stage is able to process one job at a time. Moreover, the job must be processed by exactly one machine at each stage. The purpose of the HFS scheduling problem is to determine the jobs and their processing order on each machine.

10.5.2 PROBLEM DESCRIPTION

The HFS scheduling problem considered in this study consists of S stages in series. Each stage k ($k=1,2, ..., S$) consists of i ($i=1,2, ..., M$) eligible machines in parallel. If a job is allowed to be processed on a subset of machines, these machines are called dedicated or eligible machines. A set of N independent jobs with different job types and due dates $\tilde{d}_1,...,\tilde{d}_N$ passes through S stages consecutively and is processed on any machine available at each stage. It is assumed that parallel machines at each stage are unrelated, meaning that they have a diverse speed for each job type. V_{kit} denotes the speed of machine i at stage k to process job type t. $\tilde{p}_{kt}$ indicates the processing time of job j with type t, $1 \le t \le L$ at stage k. The machine velocity represents job processing which is accomplished in a time unit. In other words, a machine with velocity V_{kit} is able to process V_{kit} units in a time unit. It means that $P_{kit} = P_{kt}/V_{kit}$ where P_{kt} is the processing time of job type t at stage k and P_{kt} is the processing time of job type t at stage k if it is processed by machine i. Completion time is the moment at which the job is completed at a stage. Completion time of job j at stage k is equal to sum of its starting time and processing time at stage k. A job processing can be started at the first stage only if the job is ready. Each job has to be assigned to exactly one machine at each stage. The objective is to minimize a linear combination of total completion time and maximum lateness. Hence, the considered problem is demonstrated

as $HFS\left|M_j\right|\alpha\sum_j \tilde{C}_j + (1-\alpha)\tilde{L}_{\max}$ and the objectives are briefly described as follows:

$\sum \tilde{C}_j$: Total job completion times at last stage

$\tilde{L}_{\max}$: Maximum lateness which is computed as follows:

$$\tilde{L}_{\max} = \max\left(\tilde{L}_j\right) = \max\left(\tilde{C}_j - \tilde{d}_j\right), j = 1,\ldots,N$$

where $\tilde{d}_j$ is the due date of job j. Assume that $w_1 = \alpha$ and $w_2 = 1-\alpha$ are weights of $\sum C_j$ and L_{max}, respectively. Consequently, final objective function is applied in the model is as follows:

$$Min\ Z = w_1\sum C_j + w_2 L_{\max}$$

The following assumptions are considered to solve the proposed HFS scheduling problem:

- The number of stages and the number of machines at each stage are known in advance.
- All jobs and machines are available at the beginning of scheduling.
- Preemption is not allowed for job processing.
- There is no breakdown for machines and they are permanently available.
- The job setup and transportation times are sequence independent and are included in the processing times.
- Parallel machines at each stage are unrelated and eligible.
- Infinite buffers exist between two consecutive stages.
- Each job is processed on one machine at a time.
- A machine cannot process more than one job simultaneously.
- Processing times and due dates are considered as fuzzy triangular numbers.
- There are different job types.
- Jobs are independent and there is no precedent constraint between them.

10.6 MATHEMATICAL MODEL

Prior to presenting the proposed model, indices, parameters and decision variables are described as follows.

10.6.1 INDICES

j: Job index, $j \in \{1, 2, \ldots, N\}$
q: Sequence index, $q \in \{1, 2, \ldots, N\}$
k: Stage index, $k \in \{1, 2, \ldots, S\}$
i: Machine index, $i \in \{1, 2, \ldots, M\}$
t: Job type index, $t \in \{1, 2, \ldots, L\}$

10.6.2 PARAMETERS

N: Total number of jobs
S: Total number of stages
M: Maximum number of machines at each stage
L: Total number of job types
e_{kit}:   If machine i at stage k is capable to process job type t, 1 otherwise 0
w_{tj}:   If job j is of type t, 1 otherwise 0
z_{ki}:   If machine i exists at stage k, 1 otherwise 0
V_{kit}:   Speed of machine i at stage k to process job type t
$\tilde{P}_{kt}$:   Fuzzy processing time of job type t at stage k
$\tilde{d}_j$:   Fuzzy due date of job j
α: The weight of objective functions
U: A positive large number.

10.6.3 DECISION VARIABLES

X_{kiqj}:   If job j at stage k is assigned to sequence q of machine i, 1 otherwise 0
$\tilde{C}_{kiqj}$:   Fuzzy completion time of job j in sequence q of machine i at stage k
$\tilde{L}_j$:   Fuzzy lateness of job j.

According to the aforementioned notations, the proposed mathematical model is presented as follows:

$$F = \min\left(\alpha F_1 + (1 - \alpha) F_2 \right) \tag{1}$$

$$F_1 = \sum_{i=1}^{M} \sum_{q=1}^{N} \sum_{j=1}^{N} \tilde{C}_{siqj} \tag{2}$$

$$F_2 = \max\left(\tilde{C}_{siqj} - \tilde{d}_j, i = 1, \ldots, M, q = 1, \ldots, N, \quad j = 1, \ldots, N \right) \tag{3}$$

s.t.

$$X_{kiqj} \leq \sum_{t=1}^{L} w_{tj} e_{kit} \qquad \forall k, i, j, q; \tag{4}$$

$$\sum_{i=1}^{M}\sum_{q=1}^{N} X_{kiqj} = 1 \qquad\qquad \forall k,j; \qquad (5)$$

$$\sum_{j=1}^{N} X_{kiqj} \leq 1 \qquad\qquad \forall k,i,q; \qquad (6)$$

$$\tilde{C}_{1i1j} \geq \sum_{t=1}^{L} X_{1i1j} \frac{\tilde{P}_{1t}}{\tilde{V}_{1it}} w_{tj} e_{1it} \qquad\qquad \forall i,j; \qquad (7)$$

$$\tilde{C}_{1iqj} + U(1 - X_{1iqj}) \geq \sum_{h \neq j} \tilde{C}_{1i(q-1)h} + \sum_{t=1}^{L} X_{1iqj} \frac{\tilde{P}_{1t}}{\tilde{V}_{1it}} w_{tj} e_{1it} \qquad \forall i,j,q; q > 1; \qquad (8)$$

$$\tilde{C}_{ki1j} + U(1 - X_{ki1j}) \geq \sum_{i'=1}^{M}\sum_{q'=1}^{N} \tilde{C}_{(k-1)i'q'j} + \sum_{t=1}^{L} X_{ki1j} \frac{\tilde{P}_{kt}}{\tilde{V}_{kit}} w_{tj} e_{kit} \qquad \forall k,i,j; k > 1; \qquad (9)$$

$$\tilde{C}_{kiqj} + U(1 - X_{kiqj}) \geq \qquad\qquad\qquad \forall k,i,j,q; \qquad (10)$$
$$\max(\sum_{i'=1}^{M}\sum_{q'=1}^{N} \tilde{C}_{(k-1)i'q'j}, \sum_{h \neq j} \tilde{C}_{ki(q-1)h}) + \sum_{t=1}^{L} X_{kiqj} \frac{\tilde{P}_{kt}}{\tilde{V}_{kit}} w_{tj} e_{kit} \qquad q > 1; k > 1;$$

$$\sum_{j \neq h} X_{kiqj} \geq X_{ki(q+1)h} \qquad\qquad \forall k,i,j,q; \qquad (11)$$

$$X_{kiqj} \in \{0,1\} \qquad\qquad \forall k,i,j,q; \qquad (12)$$

In the proposed formulations, Constraint (1) gives a linear combination of total job completion time and maximum lateness. Constraints (2) and (3) define total job completion time and maximum lateness, respectively. Constraint (4) ensures that X_{kiqj} is 1 only if job j is of type t and machine i at stage k is eligible to process it. Constraint (5) forces that each job is processed only by one machine at each stage. Constraint (6) guarantees only one job is assigned to each sequence of a machine at a moment. Constraints (7) and (8) calculate job completion times at first sequence and sequence q ($q > 1$) at the first stage, respectively. Constraints (9) and (10) calculate job completion times at first sequence and sequence q ($q > 1$) at k ($k > 1$) stages, respectively. Constraint (11) ensures that if no job is assigned to a sequence of a machine, the next sequences of the machine remain empty. Constraint (12) demonstrates binary variables.

10.6.4 Fuzzy Numbers and Calculation

In this study, processing times and due dates are considered as triangular fuzzy numbers. The Triangular Fuzzy Number (TFN) is one of the most useful fuzzy numbers which is denoted by $\tilde{a} = (a^l, a, a^u), a^l \leq a \leq a^u$. We also say that a is the core value of $\tilde{a}$ meaning that $\mu_{\tilde{a}} = (a) = 1$, and a^l and a^u are the left and right spread values of $\tilde{a}$, respectively. A TFN $\tilde{a}$ can be denoted by its membership function:

$$\mu_{\tilde{a}}(x) = \begin{cases} (x - a^l)/(a - a^l) & \text{if } a^l \leq x \leq a, \\ (a^u - x)/(a^u - a) & \text{if } a \leq x \leq a^u, \\ 0 & \text{otherwise} \end{cases}$$

In this paper, processing times and due dates are considered as $\tilde{p}_{kit} = \left(p_{kit}^L, p_{kit}, p_{kit}^U \right)$ and $\tilde{d}_j = \left(d_j^L, d_j, d_j^U \right)$, respectively. Accordingly, fuzzy completion time and lateness are proposed based on subtraction and maximum concepts of any two fuzzy numbers. The objective function is defined as the weighted sum of fuzzy lateness and fuzzy completion time based on the addition concept of fuzzy numbers. Consequently, the objective function turns into a fuzzy-valued function. Hence, in this part we explain how fuzzy completion time and fuzzy lateness are calculated. Then, the methods by which the objective function is computed and defuzzified are described.

10.6.4.1   Fuzzy Completion Time Calculation

Because processing time is a fuzzy number, namely $\tilde{p}_{kit} = \left(p_{kit}^L, p_{kit}, p_{kit}^U \right)$, by using fuzzy operands, completion time is also calculated as $\tilde{C}_{kiqj} = \left(C_{kiqj}^L, C_{kiqj}, C_{kiqj}^U \right)$. It should be noted that in completion time calculations, it is a requisite to calculate $\dfrac{\tilde{P}_{kt}}{V_{kit}}$ and since $\tilde{P}_{kt}$ is a fuzzy number, V_{kit} which is a crisp number is transformed to a fuzzy number as follows:

$$\tilde{V}_{kit} = \left(V_{kit}, V_{kit}, V_{kit} \right) = \left(V_{kit}^L, V_{kit}, V_{kit}^U \right)$$

According to the proposed explanations, fuzzy completion time is calculated as follows:

10.6.4.1.1 C_{kiqj}^L Calculation

$$C_{1i1j}^L \geq \sum_{t=1}^{L} X_{1i1j} \frac{P_{1it}^L}{V_{1it}^L} w_{tj} e_{1it} \qquad\qquad \forall i, j; \tag{13}$$

$$C_{1iqj}^L + U(1 - X_{1iqj}) \geq \sum_{h \neq j} C_{1i(q-1)h}^L + \sum_{t=1}^{L} X_{1iqj} \frac{P_{1it}^L}{V_{1it}^L} w_{tj} e_{1it} \qquad \forall i, j, q; q > 1; \tag{14}$$

$$C_{ki1j}^L + U(1 - X_{ki1j}) \geq \sum_{i'=1}^{M} \sum_{q'=1}^{N} C_{(k-1)i'q'j}^L + \sum_{t=1}^{L} X_{ki1j} \frac{P_{kt}^L}{V_{kit}^L} w_{tj} e_{kit} \qquad \forall k, i, j; k > 1; \tag{15}$$

$$C_{kiqj}^L + U(1 - X_{kiqj}) \geq \qquad\qquad\qquad \forall k, i, j, q; \tag{16}$$
$$q > 1; k > 1;$$
$$\max(\sum_{i'=1}^{M} \sum_{q'=1}^{N} C_{(k-1)i'q'j}^L, \sum_{h \neq j} C_{ki(q-1)h}^L) + \sum_{t=1}^{L} X_{kiqj} \frac{P_{kt}^L}{V_{kit}^L} w_{tj} e_{kit}$$

10.6.4.1.2 C_{kiqj} Calculation

$$C_{1i1j} \geq \sum_{t=1}^{L} X_{1i1j} \frac{P_{1it}}{V_{1it}} w_{tj} e_{1it} \qquad\qquad \forall i, j; \tag{17}$$

$$C_{1iqj} + U(1 - X_{1iqj}) \geq \sum_{h \neq j} C_{1i(q-1)h} + \sum_{t=1}^{L} X_{1iqj} \frac{P_{1it}}{V_{1it}} w_{tj} e_{1it} \qquad \forall i, j, q; q > 1; \tag{18}$$

$$C_{ki1j} + U(1 - X_{ki1j}) \geq \sum_{i'=1}^{M}\sum_{q'=1}^{N} C_{(k-1)i'q'j} + \sum_{t=1}^{L} X_{ki1j}\frac{P_{kt}}{V_{kit}}w_{tj}e_{kit} \qquad \forall k,i,j;k>1; \qquad (19)$$

$$C_{kiqj} + U(1 - X_{kiqj}) \geq \qquad\qquad\qquad\qquad \forall k,i,j,q; \qquad (20)$$
$$\max(\sum_{i'=1}^{M}\sum_{q'=1}^{N} C_{(k-1)i'q'j}, \sum_{h\neq j} C_{ki(q-1)h}) + \sum_{t=1}^{L} X_{kiqj}\frac{P_{kt}}{V_{kit}}w_{tj}e_{kit} \qquad q>1;k>1;$$

10.6.4.1.3 C_{kiqj}^{U} Calculation

$$C_{1i1j}^{U} \geq \sum_{t=1}^{L} X_{1i1j}\frac{P_{1t}^{U}}{V_{1it}^{U}}w_{tj}e_{1it} \qquad\qquad\qquad\qquad \forall i,j; \qquad (21)$$

$$C_{1iqj}^{U} + U(1 - X_{1iqj}) \geq \sum_{h\neq j} C_{1i(q-1)h}^{U} + \sum_{t=1}^{L} X_{1iqj}\frac{P_{1t}^{U}}{V_{1it}^{U}}w_{tj}e_{1it} \qquad \forall i,j,q;q>1; \qquad (22)$$

$$C_{ki1j}^{U} + U(1 - X_{ki1j}) \geq \sum_{i'=1}^{M}\sum_{q'=1}^{N} C_{(k-1)i'q'j}^{U} + \sum_{t=1}^{L} X_{ki1j}\frac{P_{kt}^{U}}{V_{kit}^{U}}w_{tj}e_{kit} \qquad \forall k,i,j;k>1; \qquad (23)$$

$$C_{kiqj}^{U} + U(1 - X_{kiqj}) \geq \qquad\qquad\qquad\qquad \forall k,i,j,q; \qquad (24)$$
$$\max(\sum_{i'=1}^{M}\sum_{q'=1}^{N} C_{(k-1)i'q'j}^{U}, \sum_{h\neq j} C_{ki(q-1)h}^{U}) + \sum_{t=1}^{L} X_{kiqj}\frac{P_{kt}^{U}}{V_{kit}^{U}}w_{tj}e_{kit} \qquad q>1;k>1;$$

According to these relations and operations, the final triangular fuzzy completion time is calculated as $\tilde{C}_{kiqj} = (C_{kiqj}^{L}, C_{kiqj}, C_{kiqj}^{U})$ for each job.

10.6.4.2 Fuzzy Lateness Calculation

Firstly, total processing times at all stages are calculated for each job:

$$\tilde{P}_j = \sum_{k=1}^{S} \tilde{P}_{kt}, \forall j \in N, t \in L.$$

Then, due date is generated as follows:

$$\tilde{d}_j = (1 + random \times 3) \times (S / \min(m_k)) \times \tilde{p}_j, \quad \forall j \in N,$$

In which *random* is a random number with a uniform distribution [0,1].

When completion time is greater than due date, there is a lateness. In conventional scheduling problems, lateness is calculated as follows:

$$L_j = C_j - d_j,$$

In which d_j, C_j and L_j are due date, completion time and lateness of job j, respectively.

In fuzzy problems, completion time $\tilde{C}_j$ and due date $\tilde{d}_j$ are considered as fuzzy numbers.

Therefore, the lateness in fuzzy environment is described as:

$$\tilde{L}_j = \tilde{C}_j \ominus \tilde{d}_j.$$

According to the aforementioned relations, $\tilde{L}_j$ is a fuzzy number and it is considered as follows:

$$\tilde{L}^L_{j\alpha} = \tilde{C}^L_{j\alpha} - \tilde{d}^U_{j\alpha}$$

$$\tilde{L}^U_{j\alpha} = \tilde{C}^U_{j\alpha} - \tilde{d}^L_{j\alpha}$$

10.6.4.2.1 Defuzzification Method

Based on the given explanations, to find the optimum amount of the final objective function, fuzzy objective functions are changed into crisp numbers. This operation, called defuzzification, is a reverse process of fuzzification and converts a fuzzy number into a crisp value. The "Center of Area" (COA) defuzzification method is a prevalent defuzzification method. It returns the center of area under the curve. Let A be a fuzzy number and μ_A as its membership function. The defuzzified amount of A by the COA method is denoted by $\delta(A)$ and returns the crisp value x^*:

$$x^* = \delta(A) = \frac{\int \mu_A(x).x dx}{\int \mu_A(x) dx}$$

For a TFN $A = (p, q, r)$, $x^* = \delta(A) = (p + q + r)/3$ and it represents the centroid of the triangle $((p, 0);(q, 1);(r, 0))$.

10.6.4.2.2 Ranking Method

Ranking of two numbers is quite important in a scheduling algorithm. It is easily performed when operands are all crisp numbers. In this study, due to some fuzzy parameters, fuzzy ranking methods are used. Several fuzzy ranking methods have been proposed so far, such as the Hamming distance method, the probability distribution method, the pseudo-order fuzzy preference model, the new fuzzy-weighted average, signed distance method, and so on. In this study, the "mean-variance" method is used. To rank two TFN $A = (a_1, a_2, a_3)$ and $B = (b_1, b_2, b_3)$, mean and variance should be calculated for both of them:

$$\bar{x}(A) = \frac{a_1 + a_2 + a_3}{3}, \quad \delta(A) = \frac{(a_1)^2 + (a_2)^2 + (a_3)^2 - (a_1 a_2 + a_1 a_3 + a_2 a_3)}{18}$$

$$\bar{x}(B) = \frac{b_1 + b_2 + b_3}{3}, \quad \delta(B) = \frac{(b_1)^2 + (b_2)^2 + (b_3)^2 - (b_1 b_2 + b_1 b_3 + b_2 b_3)}{18}$$

The TFNs are ranked by the mean-variance method as follows:

$$A > B \text{ if } \bar{x}(A) > \bar{x}(B),$$
$$A > B \text{ if } \bar{x}(A) > \bar{x}(B) \text{ and } \delta(A) < \delta(B).$$

10.6.5 FUZZY C-MEANS CLUSTERING ALGORITHM

Clustering is an efficient mechanism in data analysis to define or organize a group of patterns or objects into clusters without previous information. The objects in the same cluster share common properties and those in different clusters have distinct dissimilarity. Nowadays, due to its improvement and effectiveness, clustering has attracted a vast body of interests. Clustering plays an important role in many engineering fields such as data mining, pattern recognition, machine learning, image segmentation and so on. There are two main procedures for clustering: crisp clustering and fuzzy clustering. Crisp or hard clustering is based on classic set theory and determines whether a piece of data belongs to a cluster or not. In other words, in classic clustering each piece of input data belongs to exactly one cluster and clusters never overlap. Conversely, in fuzzy clustering, each piece of data has a different membership degree and belongs to several clusters simultaneously. However, the result of these clustering algorithms depends on input parameters. For instance, in C-means clustering algorithm, the total number of clusters (C) has to be pre-defined. Several clustering algorithms have been applied so far such as K-means, fuzzy C-means, hierarchical clustering, and so forth. The fuzzy C-means clustering algorithm is applied in this study.

The Fuzzy C-Means clustering algorithm (FCM) is one of the most reputable methods of fuzzy clustering. FCM aims to minimize an objective function with consideration of the similarity between objects and clusters' centers. The similarity a point shares with each cluster is represented with a function (called the membership function) with values (called memberships) between zero and one. Each sample has a membership in each cluster; memberships close to unity and zero signify a high and little degree of similarity respectively between the sample and a cluster. As mentioned earlier, in clustering algorithms the total number of clusters (C) has to be pre-defined. Most clustering algorithms' results depend on primary points. Therefore, clustering results may be different even if the total number of clusters is kept fixed. The optimal number of clusters, C_{opt}, is smaller or equal to the maximum number of clusters C_{max}. Although there are different methods to define the upper bound of the optimal number of clusters, a popular rule is $C_{max} \leq \sqrt{n}$.

In fuzzy clustering, the membership degrees are defined by $U = [u_{ij}]_{C \times n}$ matrix which is composed of C rows and n columns. Each u_{ij} element, $0 \leq u_{ij} \leq 1$, is the membership degree of sample x_j to cluster i. Despite the fact that the u_{ij} element is a number in [0, 1], the sum of memberships for each sample point must be unity:

$$\sum_{i=1}^{C} u_{ij} = 1, \quad \forall j = 1, ..., n, \tag{a}$$

Let $X = \{x_1, x_2, ..., x_n\}$ be a sample of n observations in p-dimensional space of R^p, namely, $X \subset R^p$ The FCM clustering algorithm aims to minimize a criterion. Several clustering criteria have been proposed to identify optimal fuzzy c-partitions in X. Here, the most popular and well-studied method is associated with the generalized least-squared errors function:

$$J(U,V,X) = \sum_{j=1}^{n} \sum_{i=1}^{C} u_{ij}^2 d_{ij}^2,$$ (b)

where,

$$d_{ij} = \sqrt{\sum_{k=1}^{s} \left(v_{ik} - x_{jk} \right)^2}$$ (c)

$V = (v_1, v_2, \ldots, v_C)$ is the vector of cluster centers with $v_i \in R^p$. Cluster centers and related membership degrees are calculated according to the following equations:

$$u_{ij} = \frac{1}{\sum_{k=1}^{C} (\frac{d_{ij}}{d_{kj}})^2}, 1 \le i \le C, 1 \le j \le n,$$ (d)

$$v_i = \frac{\sum_{j=1}^{n} u_{ij}^2 x_j}{\sum_{j=1}^{n} u_{ij}^2}, 1 \le i \le C$$ (e)

The FCM algorithm is applied as the following steps:

Step 1: Select C ($1 < C < n$) and a tolerance degree ε Initialize the membership matrix U by generating $C \times n$ random numbers in $[0,1]$.
Step 2: Compute the cluster centers v_i ($1 \le i \le c$), according to Equation (e)
Step 3: Compute d_{ij} and u_{ij} according to Equations (c) and (d), respectively. Then, update U with the new u_{ij}.
Step 4: Compute objective function J. If the difference between J and the previous objective function is less than ε stop, otherwise go to step 2.

10.7 COMPUTATIONAL RESULTS

Here, some small size problems are initially solved by GAMS software and the computation time is evaluated. The results of these numerical examples are demonstrated in Table 10.2. The computation time equals the average computation time of each problem in five runs.

According to Table 10.2, the computation time required to solve the problem with GAMS software is extremely high for small sized problems, so that no global optimal solutions were obtained for the HFS problem with 10 jobs even after 12 hours. Therefore, it is apparent that meta-heuristics are efficient to solve a complex bi-objective HFS scheduling problem with unrelated and eligible machines in fuzzy space in a reasonable amount of time.

TABLE 10.2
Computational results of GAMS for small size problems

| | Problem data | | | |
Sample	Stage number	Job number	Average run time[a]	status
1	2	3	00:00:0.2412	Optimal
2	3	3	00:00:0.6158	Optimal
3	4	3	00:00:0.4918	Optimal
4	2	4	00:00:0.6024	Optimal
5	3	4	00:00:1.278	Optimal
6	4	4	00:00:2.6852	Optimal
7	2	5	00:00:0.4108	Optimal
8	3	5	00:00:2.2886	Optimal
9	4	5	8:00:00	Local
10	2	6	00:00:1.0952	Optimal
11	3	6	00:00:16.6176	Optimal
12	4	6	12:00:00	Local
13	2	8	00:04:26.7766	Optimal
14	3	8	01:03:1.651	Optimal
15	4	8	12:00:00	Local
16	2	10	12:00:00	Local
17	3	10	12:00:00	Local
18	4	10	12:00:00	local

Note
a   Ccomputation time (hour:minute:second).

10.8   OPTIMIZATION MODEL CASE III: DELAY FLEXIBLE SCHEDULING

10.8.1   OVERVIEW OF FLEXIBLE FLOW SHOP MANUFACTURING SYSTEM

Flow shop manufacturing systems exist in many real world situations. In this system, it is assumed that the products are processed at some stages (or work centers) such that all of the products pass all of the stages in a unique sequence. The Flexible Flow Shop Manufacturing system (FFMS) is a modification of the flow shop manufacturing system in which some products do not need to be processed at some stages. This system is more general and more realistic than the flow shop system. With the popularity of the just-in-time philosophy, production managers have started to utilize this approach in their manufacturing systems. In fact, Earliness/Tardiness Production Scheduling and Planning (ETPSP) tries to implement JIT for a manufacturing system as much as possible. So every job (or product) completed earlier than its due date could cause opportunity costs for its carrying inventory until the due date, while any job not completed until its due date could cause trouble for its customer. Therefore it is desirable to get all the jobs finished as close as possible on their assigned due dates. Many scheduling research projects with Earliness/Tardiness (*E/T*) measures have been reported in the literature. This problem was originally called the minimum weighted absolute deviation problem. We developed an extensive planning approach in flexible flow shop systems in general sense. There are some stages and some jobs (or products). A total demand of each job should be delivered at the end of the planning horizon, which includes some equal periods. In other words, there is a common due date at the end of each period. The demands of each product in each period is known, so for all periods, except the last period, ETPP is determined by the weighted summation of all excess production (as the early completed products) and stockouts (as the tardy to be completed products). But, because the total demands of each job (or product) should be delivered at the end of the finite planning horizon; in the last period, ETPSP is determined by the weighted summation of earliness and tardiness time of completing all of demands of each product as compared with the final due date (the end of the last period).

10.9 PROBLEM FORMULATION

The flexible flow shop ET problem for delivering the total demand for each product during a finite planning horizon in some periods can be formulated using the following notation:

M A large number
$d_i(k)$ Demand of product i in period k
$p_i(k)$ Production quantity of product i in period k
T Number of periods within planning horizon
L Length of each demand period
L_{ek} Length of expanded period k
α_i Earliness cost per unit of product i (used for periods 1 to $T-1$)
α_i^T Earliness cost per unit of time for product i (used for period T)
E_i Earliness time of product i in the last period
β_i Tardiness cost per unit of product i (used for periods 1 to $T-1$)
β_i^T Tardiness cost per unit of time for product i (used for period T)
T_i Tardiness time of productx_i^T i in the last period
b_{ijk} Beginning time of processing product i at stage (or machine) j in period k
w_{ij} Processing time of unit of product i at stage j
s_i Lot-size of product i
l_{ik} Inventory of product i at the beginning of period k
I_{ik} Inventory of product i at the end of period k (early produced quantity)
B_{ik} Backorder of product i at the end of period k (quantity to be produced with tardiness)
N Number of products
m Number of stages.

The objective function can be expressed as follows:

$$\text{Minimize: } \sum_{i=1}^{N}\sum_{k=1}^{T-1}\left\{x_i\left[l_{ik}+\sum_{i=1}^{k}p_i(t)-\sum_{i=1}^{k}d_i(t)\right]^+ +\beta_i\left[\sum_{i=1}^{k}d_i(t)-\sum_{i=1}^{k}p_i(t)-l_{ik}\right]^+\right\}$$
$$+\sum_{i=1}^{N}\left\{x_i^T\left[LT-b_{imT}-w_{imPi}(T)\right]^+ +\beta_i^T\left[b_{imT}-w_{imPi}(T)-(LT)\right]^+\right\} \tag{1}$$

which $[x]^+$ equals zero if $x < 0$ and equals x otherwise.

As shown, Expression (1) adds the penalties of producing more and less than cumulative not responded demands until each period for periods 1 to $T-1$ and penalties originated from the difference between final due date (end of the last period) and completion time of producing not responded demands in the last period. To convert the objective function to a linear expression, we replace the Expression (1) by Expression (1′) and append eight sets of constraints ((2), (3), (4), (5), (6), (7), (8) and (9)).

$$\text{Minimize } \sum_{i=1}^{N}\sum_{k=1}^{T-1}(x_i I_{ik}+\beta_i B_{ik})+\sum_{i=1}^{N}\left\{x_i^T E_i+\beta_i^T T_i\right\}, \tag{1′}$$

Such that:

$$I_{ik} \geq I_{ik}+\sum_{i=1}^{k}p_i(t)-\sum_{i=1}^{k}d_i(t);\ \ i=1,\dots,N:k=1,\dots,T-1, \tag{2}$$

$$B_{ik} \geq \sum_{i=1}^{k}d_i(t)-\sum_{i=1}^{k}p_i(t)-I_{ik};\ \ i=1,\dots,N:k=1,\dots,T-1, \tag{3}$$

$$I_{ik} \geq 0: i=1,\dots,N;\ k=1,\dots,T-1, \tag{4}$$

$$B_{ik} \geq 0: i=1,\dots,N;\ k=1,\dots,T-1, \tag{5}$$

$$E_{ik} \geq LT-b_{imT}-w_{imPi}(T);\ \ i=1,\dots,N, \tag{6}$$

$$T_i \geq b_{ikT}+w_{imPi}(T)-LT;\ \ i=1,\dots,N, \tag{7}$$

$$E_i \geq 0;\ \ i=1,\dots,N, \tag{8}$$

$$T_i \geq 0;\ \ i=1,\dots,N. \tag{9}$$

In the above constraints, we assumed that each product batch must be processed by only one machine at any time. In addition, we assumed that each machine can process at most one product at any time. Consider two products i and i' that are to be processed at stage (or machine) j. According to the above assumption the process of product i on machine j should be started after completing the process of product i' on machine j or vice versa. To guarantee that our solution space has such a specification we add three sets of constraints (12), (13) and (14):

$$b_{i'jk} \geq b_{ijk} + w_{ijPi}(k) - My_{ii'jk}); i = 1,\ldots,N-1, i' = i+1,\ldots,N,$$
$$j = 1,\ldots,m \left| w_{ij} \neq 0 \ \& \ w_{ij} \neq 0, k = 1,\ldots,T, \right. \tag{12}$$

$$b_{ijk} \geq b_{i'jk} + w_{i'jPi'}(k) - M(1 - y_{ii'jk}); i = 1,\ldots,N-1, i' = i+1,\ldots,N,$$
$$j = 1,\ldots,m \left| w_{ij} \neq 0 \ \& \ w_{ij} \neq 0, k = 1,\ldots,T, \right. \tag{13}$$

$$Y_{ii'jk} \in \{0,1\}; i = 1,\ldots,N-1, i' = 1,\ldots,N \ j = 1,\ldots,m \left| w_{ij} \neq 0 \ \& \ w_{i'j} \neq 0, k = 1,\ldots,T. \right. \tag{14}$$

It is obvious that the production of each period on each machine should be started after that the production of its last period is finished, so we append the below constraints to the model:

$$b_{ijk} \geq b_{i'jk-1} + w_{i'jPi'}(k-1); i = 1,\ldots,N, i' = 1,\ldots,N,$$
$$j = 1,\ldots,m \left| w_{ij} \neq 0 \ \& \ w_{i'j} \neq 0, k = 2,\ldots,T. \right. \tag{15}$$

In many practical environments, operation managers consider lot-size for each product. The production quantity of each item should be a product of its lot-size. we consider this characteristic in our problem by adding the sets of constraints (16) and (17)

$$p_i(k) = r_i s_i; \ i = 1,\ldots,N, k = 1,\ldots,T-1, \tag{16}$$

$$r_i \in \{0,1,2,\ldots\}; i = 1,\ldots,N. \tag{17}$$

l_{i1} is known as parameter, but l_{ik} for periods 2 to $T - 1$ is calculated by the set of constraints (18)

$$l_{ik} = l_{ik-1} + p_i(k-1) - d_i(k-1); i = 1,\ldots,N, k = 2,\ldots,T-1. \tag{18}$$

However, since we should deliver the total demands of all periods of the planning horizon, the production quantity of each product is not a decision variable anymore in the last period. In fact, the quantity of production for product i in the last period is calculated by Equation (19) as a set of constraints

$$p_i(T) = d_{iT} - l_{iT}; i = 1,\ldots,N. \tag{19}$$

10.10   NUMERICAL EXAMPLE

A small firm has one production line including four stages. This firm has contracted to supply four parts for an assembling company within next six weeks. The processing time of each part on each machine expressed as follows in Table 10.3.

Lot sizes of the parts 1–4 are respectively 50, 100, 50, 100. The demand of each part at the end of each week expressed in Table 10.4.

Each week includes 6 workdays and each workday, in turn, includes 480 minutes, that is 2880 min (excluding idle times). There is no on hand inventory of each part at the beginning of the planning horizon. The earliness/tardiness penalties of each part are expressed in Table 10.5.

After solving the model for period 1, the optimal production quantities for this subproblem are:

TABLE 10.3
Processing time of each unit part on each machine (min)

Product	Operation 1	Operation 2	Operation 3	Operation 4
1	2.1	1.8	–	1.6
2	–	1.7	1.5	1.8
3	–	2.1	2.3	–
4	1.8	–	1.7	2.0

TABLE 10.4
Demand of each product (part) to be delivered at the end of each week

Part	Period 1	2	3	4	5	6	Total
1	400	350	600	300	600	600	2850
2	450	300	350	450	300	550	2400
3	550	450	400	400	600	600	3000
4	350	400	600	300	500	550	2700

TABLE 10.5
E/T penalties of each part ($)

Part (i)	α_i	x_i^T	β_i	β_i^T
1	2	10	4	10
2	3	10	5	15
3	2	10	3	10
4	3	10	4	12

$$p_1(1) = 400; p_2(1) = 400; p_3(1) = 400; p_4(1) = 300.$$

The optimal value of the objective function in this period equals 900\$. We can conclude that in period 2:

$L = 2880 + (2880 + 1520) = 4240\,\text{min},$

$b_{il2} \geq 0; i = 1,\ldots,4,$
$b_{i22} \geq 2240 - 1520; i = 1,\ldots,4,$
$b_{i32} \geq 2570 - 1520; i = 1,\ldots,4,$
$b_{i42} \geq 2880 - 1520; i = 1,\ldots,4.$

Furthermore, initial inventory of each part in period 2 is:

$$l_{12} = 0; \ l_{22} = -50; \ l_{32} = -150; \ l_{42} = -50.$$

After solving the model for period 2, the optimal production quantities for this subproblem are:

$$p_1(2) = 350; \ p_2(2) = 400; \ p_3(2) = 600; \ p_4(2) = 500.$$

The optimal value of the objective function in this period equals 300\$. We can conclude that in period 3:

$L = 2880 + (2880 - 1690) = 4070\,\text{min},$
$b_{il3} \geq 0; \ i = 1,\ldots,4,$
$b_{i23} \geq 2320 - 1690; \ i = 1,\ldots,4,$
$b_{i33} \geq 2880 - 1690; \ i = 1,\ldots,4,$
$b_{i42} \geq 2880 - 1690; \ i = 1,\ldots,4.$

Furthermore, initial inventory of each part in period 3 is:

$l_{13} = 0; \ l_{23} = 50; \ l_{23} = 0; \ l_{43} = 50.$

So, the optimal production quantities for period 3 are:

$$p_1(3) = 500; \ p_2(3) = 400; \ p_3(3) = 400; \ p_4(3) = 500.$$

The optimal value of the objective function in this period equals 1000\$. We can conclude that in period 4:

$$L = 2880 + (2880 - 1010) = 4750\,\text{min},$$
$$b_{il4} \geq 0; \ i = 1,\ldots,4,$$
$$b_{i24} \geq 1860 - 1010; \ i = 1,\ldots,4,$$
$$b_{i34} \geq 2780 - 1010; \ i = 1,\ldots,4,$$
$$b_{i44} \geq 2860 - 1010; \ i = 1,\ldots,4.$$

Furthermore, initial inventory of each part in period 4 is:

$$l_{14} = -100; \ l_{24} = 0; \ l_{34} = 0; \ l_{44} = -150.$$

So, the optimal production quantities for period 4 are:

$$p_1(4) = 400; \ \mathrm{p}_2(4) = 500; \ \mathrm{p}_3(4) = 400; \ p_4(4) = 500.$$

The optimal value of the objective function in this period equals 300$. So, for period 5:

$$
\begin{aligned}
L &= 2880 + (2880 - 0) = 5760 \, \text{min}, \\
b_{i15} &\geq 0; \ i = 1,\dots,4, \\
b_{i25} &\geq 1750 - 0; \ i = 1,\dots,4, \\
b_{i35} &\geq 2670 - 0; \ i = 1,\dots,4, \\
b_{i44} &\geq 2790 - 0; \ i = 1,\dots,4.
\end{aligned}
$$

Furthermore, initial inventory of each part in period 5 is:

$$l_{15} = 0; \ l_{25} = 50; \ l_{35} = 0; \ l_{45} = 50.$$

The optimal value of the objective function in the last period equals 256.67$.

10.11 CONCLUSIONS

In this chapter, three optimization cases were discussed and illustrated. The first section presented the robotic flexible flow shop with sequence-dependent setup time and an AGV-based transportation system is considered, in which the machines in stages can be both related and unrelated. There are two proposed patterns for AGVs movements, by which AGVs encounters are prevented, and the objective function is to minimize the makespan. A Mixed Integer Linear Programming (MILP) model was suggested for the problem. To evaluate the performance of the proposed model, a numerical example was handled using CPLEX. For future research, in order to prevent AGVs from being encountered, it is recommended to use two separate routes for AGVs (forward and backward). Therefore, the constraints to prevent the AGVs from facing one another in one route between each pair of machines are suggested.

The second section considered the HFS scheduling problem with unrelated and eligible machines along with fuzzy processing times and fuzzy due dates. A fuzzy mixed integer mathematical model was presented in order to minimize a linear combination of total completion time and maximum lateness of jobs. As the size of the problem and its complexity grew, the search space expanded and the computation time was increased exponentially. As the results indicated, when the problem size increases, the difference between optimal solutions resulting from clustering and conventional algorithms would be more significant. As a direction for future research, it would be interesting to consider transportation time between stages or preventive maintenance for machines. Moreover, it is worthwhile to seek a dynamic model of the problem where there are precedence constraints, aging effects and so on. Besides, other optimization heuristic and meta-heuristic algorithms may be investigated for the problem.

In the third section, we presented an extensive method for the production planning and scheduling with E/T measures in flexible manufacturing systems. We considered a finite planning horizon that is divided into equal periods. The E/T penalty was considered quantity-based during all periods except the last one. In the last period the E/T measure is changed to a time-based function because the production quantities were not decision variable in this period anymore.

BIBLIOGRAPHY

Bagchi, U., Cheng, Y.L., & Sullivan, R.S. (1987). Minimizing absolute and squared deviations of completion times with different earliness and tardiness penalties and a common due date. *Naval Research Logistics*, 34, 739–751

Balin, S. (2012). Non-identical parallel machine scheduling with fuzzy processing times using genetic algorithm and simulation. *International Journal of Advanced Manufacturing Technology*, 61(9–12), 1115–1127.

Batur, G.D., Karasan, O.E., Akturk, M.S. (2010). Multiple part-type scheduling in flexible robotic cells. *International Journal of Production Research*, 135(2), 726–740.

Behnamian, J., & Zandieh, M. (2011). A discrete colonial competitive algorithm for hybrid flow shop scheduling to minimize earliness and quadratic tardiness penalties. *Expert System with Applications*, 38, 14490–14498.

Behnamian, J., & Zandieh, M. (2013). Earliness and tardiness minimizing on a realistic hybrid flow shop scheduling with learning effect by advanced metaheuristic. *Arabian Journal for Science & Engineering*, 38(5), 1229–1242.

Benamian, J., & Fatemi Ghomi, S.M.T. (2010). Development of a PSO-SA hybrid meta-heuristic for a new comprehensive regression model to time-series forecasting. *Expert System with Applications*, 37, 974–984.

Bezdek, J., Ehrlich, R., & Full, W. (1984). FCM: The fuzzy c-means clustering algorithm. *Computers and Geosciences*, 10(2–3), 191–203.

Carlier, J., Haouari, M., Kharbeche, M., et al. (2010). An optimization-based heuristic for the robotic cell problem. *European Journal of Operational Research*, 202, 636–645.

Cheng, R.W. Gen, M.S., & Tozawa, T. (1995). Minmax earliness/tardiness scheduling in identical parallel machine system using genetic algorithms, *Computers and Industrial Engineering*, 29(1–4), 513–517.

Eddaly, M., Jarboui, B., & Siarry, P. (2016). Combinatorial particle swarm optimization for solving blocking flowshop scheduling problem. *Journal of Computational Design and Engineering*, 3(4), 295–311.

Elmi, A., & Topaloglu, S. (2013). A scheduling problem in blocking hybrid flow shop robotic cells with multiple robots. *Computers & Operations Research* 40(10), 2543–2555.

Engin, O., Ceran, G., & Yilmaz, M.K. (2011). An efficient genetic algorithm for hybrid flow shop scheduling with multiprocessor task problems. *Applied Soft Computing*, 11, 3056-3065.

Fazlollahtabar, H. (2016). Parallel autonomous guided vehicle assembly line for a semi-continuous manufacturing system. *Assembly Automation*, 36(3), 262–273.

Fazlollahtabar, H. (2018a). Lagrangian relaxation method for optimizing delay of multiple autonomous guided vehicles. *Transportation Letters*, 10(6), 354–360.

Fazlollahtabar, H. (2018b). Scheduling of multiple autonomous guided vehicles for an assembly line using minimum cost network flow. *Journal of Optimization in Industrial Engineering*, 11(1), 185–193.

Fazlollahtabar, H. (2019a). An effective mathematical programming model for production of automatic robot path planning. *The Open Transportation Journal*, 13(1), 1–16.

Fazlollahtabar, H. (2019b). Triple state reliability measurement for a complex autonomous robot system based on extended triangular distribution. *Measurement*, 139, 122–126.

Fazlollahtabar, H. (2020). Comparative simulation study for configuring turning point in multiple robot path planning: Robust data envelopment analysis. *Robotica*, 38(5), 925–939.

Fazlollahtabar, H. (2021). Robotic manufacturing systems using Internet of Things: New Era of facing pandemics. *Automation, Robotics & Communications for Industry*, 4.0, 82.

Fazlollahtabar, H. (2022). Internet of Things-based SCADA system for configuring/reconfiguring an autonomous assembly process. *Robotica*, 40(3), 672–689.

Fazlollahtabar, H., & Hassanli, S. (2018). Hybrid cost and time path planning for multiple autonomous guided vehicles. *Applied Intelligence*, 48, 482–498.

Fazlollahtabar, H., & Jalali, S.G. (2013). Adapted Markovian model to control reliability assessment in multiple AGV. *Scientia Iranica*, 20(6), 2224–2237.

Fazlollahtabar, H., & Niaki, S.T.A. (2017a). Binary state reliability computation for a complex system based on extended Bernoulli trials: Multiple autonomous robots. *Quality and Reliability Engineering International*, 33(8), 1709–1718.

Fazlollahtabar, H., & Niaki, S.T.A. (2017b). Integration of fault tree analysis, reliability block diagram and hazard decision tree for industrial robot reliability evaluation. *Industrial Robot: An International Journal*, 44(6), 754–764.

Fazlollahtabar, H., & Niaki, S.T.A. (2017c). *Reliability Models of Complex Systems for Robots and Automation*. CRC Press.

Fazlollahtabar, H., & Niaki, S.T.A. (2018a). Cold standby renewal process integrated with environmental factor effects for reliability evaluation of multiple autonomous robot system. *International Journal of Quality & Reliability Management*, 35(10), 2450–2464.

Fazlollahtabar, H., & Niaki, S.T.A. (2018b). Modified branching process for the reliability analysis of complex systems: Multiple-robot systems. *Communications in Statistics-Theory and Methods*, 47(7), 1641–1652.

Fazlollahtabar, H., & Saidi-Mehrabad, M. (2015a). *Autonomous Guided Vehicles: Methods and Models for Optimal Path Planning*. Germany: Springer International Publishing.

Fazlollahtabar, H., & Saidi-Mehrabad, M. (2015b). Risk assessment for multiple automated guided vehicle manufacturing network. *Robotics and Autonomous Systems*, 74, 175–183.

Fazlollahtabar, H., & Saidi-Mehrabad, M. (2019). *Cost Engineering and Pricing in Autonomous Manufacturing Systems*. Emerald Publishing Limited.

Fazlollahtabar, H., Mahdavi-Amiri, N., & Muhammadzadeh, A. (2015). A genetic optimization algorithm for nonlinear stochastic programs in an automated manufacturing system. *Journal of Intelligent & Fuzzy Systems*, 28(3), 1461–1475.

Fazlollahtabar, H., Saidi-Mehrabad, M., & Balakrishnan, J. (2015a). Mathematical optimization for earliness/tardiness minimization in a multiple automated guided vehicle manufacturing system via integrated heuristic algorithms. *Robotics and Autonomous Systems*, 72, 131–138.

Fazlollahtabar, H., Saidi-Mehrabad, M., & Balakrishnan, J. (2015b). Integrated Markov-neural reliability computation method: A case for multiple automated guided vehicle system. *Reliability Engineering & System Safety*, 135, 34–44.

Fazlollahtabar, H., Saidi-Mehrabad, M., & Masehian, E. (2015). Mathematical model for dead-lock resolution in multiple AGV scheduling and routing network: a case study. *Industrial Robot: An International Journal*, 42(3), 252–263.

Fazlollahtabar, H., Saidi-Mehrabad, M., & Masehian, E. (2021). Robotic industrial automation simulation-optimization for resolving conflict and deadlock. *Assembly Automation*, 41(4), 477–485.

Fazlollahtabar, H., & Shafieian, S.H. (2014). An optimal path in an AGV-based manufacturing system with intelligent agents. *Journal for Manufacturing Science and Production*, 14(2), 87–102.

Graham, R. L., Lawler, E. L., Lenstra, J. K., Rimnooy Kan, A. H. G. (1979) Optimization and Approximation in Deterministic Sequencing and Scheduling: a Survey, Editors: Hammer, P. L., Johnson, E. L., Korte, G. H., Annals of Discrete Mathematics, Elsevier, Volume 5, Pages 287-326.

Guanlong, D., Zhenhao, X., & Xingsheng, G. (2012). A Discrete artificial bee colony algorithm for minimizing the total flow time in the blocking flow shop scheduling. *Chinese Journal of Chemical Engineering*, 20, 1067–1073.

Hekmatfar, M., Fatemi Ghomi, S.M.T., & Karimi, B. (2011). Two stage reentrant hybrid flow shop with setup times and the criterion of minimizing makespan. *Applied Soft Computing*, 9, 4530–4539.

Inna, G., Drobouchevitch, H., Geismar, H.N., et al (2010). Throughput optimization in robotic cells with input and output machine buffers: A comparative study of two key models. *European Journal of Operational Research*, 206(3), 623–633.

Jabbarizadeh, F., Zandieh, F.M., & Talebi, D. (2009). Hybrid flexible flow shops with sequence-dependent setup times and machines availability constraints. *Computers and Industrial Engineering*, 57(3), 949–957.

Johnson, S.M, 1954, Optimal two–and three-stage production schedules with setup times included, *Naval Research Logistics Quarterly*, vol. 1, No. 1, pp. 61–80.

Kalczynski, P.J., & Kamburowski, J. (2012). An empirical analysis of heuristics for solving the two machine flow shop problem with job release times. *Computers & Operations Research*, 39, 2659–2665.

Khalouli, S., Ghedjati, F., & Hamzaoui, A. (2010). A meta-heuristic approach to solve a JIT scheduling problem in a hybrid flow shop. *Engineering Applications of Artificial Intelligence*, 23(5), 765–771.

Kharbeche, M., Carlier, J., Haouari, M., et al (2010). Exact Method for Robotic Cell Problem. *Electronic Notes in Discrete Mathematics*, 36, 859–866.

Lee, C. Y., & Chen, Z. L. (2001) Machine Scheduling with transportation considerations, Journal of Scheduling, Vol. 4, pp. 3-24.

Lei, D. (2015). Variable neighborhood search for two-agent flow shop scheduling problem. *Computer & Industrial Engineering*, 80, 125–131.

Li, Z.T., Liu, J., Chen, Q.X., et al (2015). Approximation algorithms for the three-stage flexible flow shop problem with mid group constraint. *Expert System with Applications*, 42, 3571–3584.

Liu Min, Wu Cheng, (2006). Genetic algorithms for the optimal common due date assignment and the optimal scheduling policy in parallel machine earliness/tardiness scheduling problems, *Robotic and Computer-Integrated Manufacturing*, 22(4), 279–287.

Low, C.Y, Hsu, C.J, & Su, C.T. (2008). A two-stage hybrid flow shop scheduling problem with a function constraint and unrelated alternative machines. *Computers & Operations Research*, 35, 845–853.

M, MK., Tosun, O., & Geetha, M. (2017). Hybrid monkey search algorithm for flow shop scheduling problem under makespan and total flow time. *Applied Soft Computing*, 55, 82–92.

Ahmed, M.U., & Sundararaghavan, P.S. (1990). Minimizing the weighted sum of late and early completion penalties in a single machine. *IIE Transactions*, 22(3), 288–290.

Mahdavi, I., Zarezadeh, V., & Shahnazari-Shahrezaei, P. (2011). Flexible flowshop scheduling with equal number of unrelated parallel machines, *Journal of Industrial Engineering International*, 7(13), 74–83.

Marichelvam, M.K., Prabaharan, T., & Yang, X.S. (2014). Improved cuckoo search algorithm for hybrid flow shop scheduling problems to minimize makespan. *Applied Soft Computing*, 19, 93–101.

Marichelvam, M.K., Prabaharan, T., & Yang, XX. (2014). Improved cuckoo search algorithm for hybrid flow shop scheduling problems to minimize makespan, *Applied Soft Computing*, 19, 93–101.

Moslehi, G., & Khorasanian, D. (2013). Optimizing blocking flow shop scheduling problem with total completion time criterion. *Computers & Operations Research*, 40, 1874–1883.

Ruiz, R., & Vazquez-Rodriguez, J.A. (2010). The hybrid flow shop scheduling problem. *European Journal of Operational Research*, 205(1), 1–18.

Sangsawang, C., Sethanan, K., Fujimoto, T., et al 2015. Metaheuristics optimization approaches for two-stage reentrant flexible flow shop with blocking constraint. *Expert System with Applications*, 42, 2395–2410.

Sawik, T. (2012). batch versus cyclic scheduling of flexible flow shops by mixed-integer programming. *International Journal of Production Research*, 50, 5017–5034.

Seidgar, H., Abedi, M., Tadayonirad, S., et al (2015). A hybrid particle swarm optimisation for scheduling just-in-time single machine with preemption, machine idle time and unequal release times. *International Journal of Production Research*, 53(6), 1912–1935.

Shojaeifar, A., Fazlollahtabar, H., & Mahdavi, I. (2016). Decomposition versus minimal path and cuts methods for reliability evaluation of an advanced robotic production system. *Journal of Automation Mobile Robotics and Intelligent Systems*, 10(3), 52–57.

Tavakkoli-Moghaddam, R., Javadi, B., & Jolai, F., et al (2010). The use of a fuzzy multiobjective linear programming for solving a multi objective single-machine scheduling problem, *Applied Soft Computing Journal*, vol. 10, no. 3, pp. 919–925.

Wang, L., Pan, Q.K., & Fatih Tasgetiren, M. (2011). A hybrid harmony search algorithm for the blocking permutation flow shop scheduling problem. *Computers & Industrial Engineering*, 61, 76–83.

Wang, S., & Liu, M. (2014). Two-stage hybrid flow shop scheduling with preventive maintenance using multi-objective tabu search method. *International Journal of Production Research*, 52(5), 1495–1508.

Warren Liao, T, & Su, P. (2017). Parallel machine scheduling in fuzzy environment with hybrid ant colony optimization including a comparison of fuzzy number ranking methods in consideration of spread of fuzziness. *Applied Soft Computing*, 56, 65–81.

Woo, Y.B, Jung, S, & Kim, B.S. (2017). A rule-based genetic algorithm with an improvement heuristic for unrelated parallel machine scheduling problem with time-dependent deterioration and multiple rate-modifying activities, *Computers & Industrial Engineering*, 109, 179–190.

Yang, J. (2015). Minimizing total completion time in a two-stage hybrid flow shop with dedicated machines at the first stage. *Computers & Operations Research*, 58, 1–18.

Yeung, W.K., Ceyda Ôgaz, & Edwin Cheng, T.C. (2004). Two-stage flowshop earliness and tardiness machine scheduling involving a common due window. *International Journal of Production Economics*, Volume 90, Issue 3, 421–434.

Zadeh, L.A. (1965). Fuzzy sets. *Information and Control*, 8, 338–353.

Zandieh, M, Fatemi Ghomi, S.M.T, & Moattar Husseini, S.M. (2006). An immune algorithm approach to hybrid flow shops scheduling with sequence-dependent setup times. *Applied Mathematics and Computation*, 180(1), 111–127.

Zhiwei Zhu, Heady, Ronald B. (2000). Minimizing the sum of earliness/tardiness in multi-machine scheduling: A mixed integer programming approach. *Computers and Industrial Engineering*, 38, 297–305.

Ziaeifar, A, Tavakkoli-Moghaddam, R, & Pichka, K. (2012). Solving a new mathematical model for a hybrid flow shop scheduling problem with a processor assignment by a genetic algorithm. *International Journal of Advanced Manufacturing Technology*, 61(1–4), 339–349.

Bibliography

Fazlollahtabar, H. (2016). Parallel autonomous guided vehicle assembly line for a semi-continuous manufacturing system. *Assembly Automation*, 36(3), 262–273.

Fazlollahtabar, H. (2018a). Lagrangian relaxation method for optimizing delay of multiple autonomous guided vehicles. *Transportation Letters*, 10(6), 354–360.

Fazlollahtabar, H. (2018b). Scheduling of multiple autonomous guided vehicles for an assembly line using minimum cost network flow. *Journal of Optimization in Industrial Engineering*, 11(1), 185–193.

Fazlollahtabar, H. (2019a). An effective mathematical programming model for production of automatic robot path planning. *The Open Transportation Journal*, 13(1), 11–16.

Fazlollahtabar, H. (2019b). Triple state reliability measurement for a complex autonomous robot system based on extended triangular distribution. *Measurement*, 139, 122–126.

Fazlollahtabar, H. (2020). Comparative simulation study for configuring turning point in multiple robot path planning: Robust data envelopment analysis. *Robotica*, 38(5), 925–939.

Fazlollahtabar, H. (2021). Robotic Manufacturing Systems Using Internet of Things: New Era of Facing Pandemics. *Automation, Robotics & Communications for Industry* 4.0, 82.

Fazlollahtabar, H. (2022). Internet of Things-based SCADA system for configuring/reconfiguring an autonomous assembly process. *Robotica*, 40(3), 672–689.

Fazlollahtabar, H., & Mahdavi-Amiri, N. (2013). Design of a neuro-fuzzy–regression expert system to estimate cost in a flexible job shop automated manufacturing system. *The International Journal of Advanced Manufacturing Technology*, 67, 1809–1823.

Fazlollahtabar, H., & Mahdavi-Amiri, N. (2013). Producer's behavior analysis in an uncertain bicriteria AGV-based flexible job shop manufacturing system with expert system. *The International Journal of Advanced Manufacturing Technology*, 65, 1605–1618.

Fazlollahtabar, H., & Mahdavi-Amiri, N. (2013). An optimal path in a bi-criteria AGV-based flexible job shop manufacturing system having uncertain parameters. *International Journal of Industrial and Systems Engineering*, 13(1), 27–55.

Fazlollahtabar, H., & Olya, M.H. (2013). A cross-entropy heuristic statistical modeling for determining total stochastic material handling time. *The International Journal of Advanced Manufacturing Technology*, 67, 1631–1641.

Fazlollahtabar, H., & Shafieian, S.H. (2014). An Optimal Path in an AGV-based manufacturing system with intelligent agents. *Journal of Manufacturing Science and Production*, 14(2), 87–102.

Fazlollahtabar, H., & Hassanli, S. (2018). Hybrid cost and time path planning for multiple autonomous guided vehicles. *Applied Intelligence*, 48, 482–498.

Fazlollahtabar, H., & Jalali, S.G. (2013). Adapted Markovian model to control reliability assessment in multiple AGV. *Scientia Iranica*, 20(6), 2224–2237.

Fazlollahtabar, H., & Niaki, S.T.A. (2017). Binary state reliability computation for a complex system based on extended Bernoulli trials: Multiple autonomous robots. *Quality and Reliability Engineering International*, 33(8), 1709–1718.

Fazlollahtabar, H., & Niaki, S.T.A. (2017). Integration of fault tree analysis, reliability block diagram and hazard decision tree for industrial robot reliability evaluation. *Industrial Robot: An International Journal*, 44(6), 754–764.

Fazlollahtabar, H., & Niaki, S.T.A. (2017). *Reliability Models of Complex Systems for Robots and Automation*. CRC Press.

Fazlollahtabar, H., & Niaki, S.T.A. (2018). Cold standby renewal process integrated with environmental factor effects for reliability evaluation of multiple autonomous robot system. *International Journal of Quality & Reliability Management*, 35(10), 2450–2464.

Fazlollahtabar, H., & Niaki, S.T.A. (2018). Modified branching process for the reliability analysis of complex systems: Multiple-robot systems. *Communications in Statistics-Theory and Methods*, 47(7), 1641–1652.

Fazlollahtabar, H., & Saidi—Mehrabad, M. (2015). Risk assessment for multiple automated guided vehicle manufacturing network. *Robotics and Autonomous Systems*, 74, 175–183.

Fazlollahtabar, H., & Saidi-Mehrabad, M. (2019). *Cost Engineering and Pricing in Autonomous Manufacturing Systems*. Emerald Publishing Limited.

Fazlollahtabar, H., & Shafieian, S.H. (2014). An optimal path in an AGV-based manufacturing system with intelligent agents. *Journal for Manufacturing Science and Production*, 14(2), 87–102.

Fazlollahtabar, H., Es'haghzadeh, A., Hajmohammadi, H., & Taheri-Ahangar, A. (2012). A Monte Carlo simulation to estimate TAGV production time in a stochastic flexible automated manufacturing system: a case study. *International Journal of Industrial and Systems Engineering*, 12(3), 243–258.

Fazlollahtabar, H., Mahdavi-Amiri, N., & Muhammadzadeh, A. (2015). A genetic optimization algorithm for nonlinear stochastic programs in an automated manufacturing system. *Journal of Intelligent and Fuzzy Systems*, 28(3), 1461–1475.

Fazlollahtabar, H., Rezaie, B., & Kalantari, H. (2010). Mathematical programming approach to optimize material flow in an AGV-based flexible job shop manufacturing system with performance analysis. *The International Journal of Advanced Manufacturing Technology*, 51(9–12), 1149–1158.

Fazlollahtabar, H., Saidi-Mehrabad, M., & Balakrishnan, J. (2015). Mathematical optimization for earliness/tardiness minimization in a multiple automated guided vehicle manufacturing system via integrated heuristic algorithms. *Robotics and Autonomous Systems*, 72, 131–138.

Fazlollahtabar, H., Saidi-Mehrabad, M., & Balakrishnan, J. (2015). Integrated Markov-neural reliability computation method: A case for multiple automated guided vehicle system. *Reliability Engineering & System Safety*, 135, 34–44.

Fazlollahtabar, H., Saidi-Mehrabad, M., & Masehian, E. (2015). Mathematical model for deadlock resolution in multiple AGV scheduling and routing network: A case study. *Industrial Robot: An International Journal*, 42(3), 252–263.

Fazlollahtabar, H., Saidi-Mehrabad, M., & Masehian, E. (2021). Robotic industrial automation simulation-optimization for resolving conflict and deadlock. *Assembly Automation*, 41(4), 477–485.

Fazlollahtabar, H., & Saidi-Mehrabad, M. (2015). *Autonomous Guided Vehicles: Methods and Models for Optimal Path Planning*. Germany: Springer International Publishing.

Shirazi, B., Fazlollahtabar, H., & Mahdavi, I. (2010). A six sigma based multi-objective optimization for machine grouping control in flexible cellular manufacturing systems with guide path flexibility. *Advances in Engineering Software*, 41(6), 865–873.

Shojaeifar, A., Fazlollahtabar, H., & Mahdavi, I. (2016). Decomposition versus Minimal Path and Cuts Methods for Reliability Evaluation of an Advanced Robotic Production System. *Journal of Automation Mobile Robotics and Intelligent Systems*, 10(3), 52–57.

Tavana, M., Fazlollahtabar, H., & Hassanzade, R. (2014). A bi-objective stochastic programming model for optimizing automated material handling systems with reliability considerations. *International Journal of Production Research* 52(19), 5597–5610.

Index